CELL RESPIRATION

W. O. JAMES, F.R.S.
Fellow of Imperial College, London

 THE ENGLISH UNIVERSITIES PRESS LTD

Modern Biology Series

General Editor
J. E. Webb, Ph.D., D.Sc.
Professor of Zoology,
Westfield College, University of London

ISBN 0 340 14800 4   Boards
ISBN 0 340 14801 2   Paperback

First printed 1971

The English Universities Press Ltd
St Paul's House, Warwick Lane, London, EC4P 4AH

Printed and bound in Great Britain by
Fletcher & Son Ltd, Norwich.

Preface

CELL respiration is a basal biological study and in many respects it is unashamedly academic. Nevertheless, nothing in cells happens without the support of their respiration, from the division that first brings them into existence through the maintenance of their maturity and their decline into senescence. The contribution of respiratory studies to useful knowledge is often therefore at second or third hand; but to some technologies it has been direct enough; as for example in its effects on brewing, fruit storage, explosives and, of course, medicine. The following pages seek to give as simple and direct an account as may interest the student embarking on its investigation.

In a short account of a subject which cuts across nearly all the conventional boundaries of biological teaching, it is not possible to detail current controversies; but, among the suggestions for further reading, there are included some which will enable those whose interest is aroused to witness the conflicts at the present frontiers of knowledge.

The preparation of such a book as this inevitably places a heavy strain upon the goodwill and patience of one's colleagues. Particularly, I would like to thank Professor C. P. Whittingham and Dr J. W. Millbank for throwing light into some dark places for me; and Mr A. D. Greenwood, Dr J. Gay and Dr H. B. Griffiths of this college; Professor E. G. Gray and Mr R. A. Willis of University College, London, and Dr H. Moor of the Eidgenössische Technische Hochschule, Zürich, for supplying original electronmicrograms. Special thanks are due to my wife for so carefully editing the manuscript and to Mrs J. Cheston for preparing it.

I also wish to express my gratitude to the following authors and publishers for permission to reproduce the following:

J. Baddiley and *Endeavour* (Plate I*a*); J. E. Roth, C. W. Lewis and R. P. Williams and the American Society for Microbiology (Plate I*b*); D. M. Judge and M. S. Anderson and the American Society for Parasitologists (Plate V*a*); K. Vickerman and F. E. Cox and John Murray (Plate V*b*); K. R. Porter and M. A. Bonneville and Henry Kimpton (Plate XII*a*).

Abbreviations and Symbols

Å	Ångstrom unit, 10^{-10} metre	μm	micron, 10^{-6} metre, (μ)
AA	ascorbic acid	MW	molecular weight
ADP	adenosine diphosphate	NAD	nicotinamide adenine dinucleotide, (DPN, CoI)
AMP	adenosine monophosphate, adenylic acid		
ATP	adenosine triphosphate	$NADH_2$	(or $NADH + H^+$) reduced nicotinamide adenine dinucleotide
BAL	British anti-Lewisite	NADP	nicotinamide adenine dinucleotide phosphate (TPN, CoII)
CoA	coenzyme A, 3-phospho-adenosinediphosphate pantetheine		
CoASH	reduced CoA	$NADPH_2$	(or $NADPH + H^+$) reduced nicotinamide adenine dinucleotide
CoQ	coenzyme Q, ubiquinone		
dh	dehydrogenase	nm	nanometer, 10^{-9} metre, (mμ)
DHA	dehydroascorbic acid		
DIECA	diethyldithiocarbamate	(P)	phosphate, $-H_2PO_3$
DNA	deoxyribonucleic acid	~(P)	phosphate with 'energy rich' linkage
DNP	2,4-dinitrophenol		
e	electron	P_i	'inorganic orthophosphate', acid or salt
E	redox potential		
E_0	standard redox potential		
E_0'	standard redox potential at pH 7	PEP	phosphoenolpyruvic acid
FAD	flavin adenine dinucleotide	PP_i	'inorganic pyrophosphate', acid or salt
FMN	flavin mononucleotide		
FP	flavoprotein	PPP	pentosephosphate pathway, hexosemono-phosphate shunt
ΔG	free energy change		
ΔG_0	standard free energy change		
$\Delta G_0'$	standard free energy change at pH 7	Q_{CO_2}	μl $_{CO_2}$/h mg dry weight
		Q_{O_2}	μl $_{O_2}$/h mg dry weight
GTP	guanosine triphosphate	RNA	ribonucleic acid
HdP	hexosediphosphate, fructofuranose-1,6-diphosphate	SCoA	oxidised CoA
		TCA	tricarboxylic acid cycle, citric acid cycle, Krebs cycle
HO(P)	phosphoric acid, H_3PO_4	TPP	thiamine pyrophosphate, diphosphothiamine
lip S_2	α-lipoic acid, 6, 8-dithio-n-octanoic acid		
lip SH_2	dihydrolipoic acid	UDPG	uridine diphosphoglucose

Contents

1 The System

ALL that lives, grows and ages. This is not a thing that starts at some particular stage of the life history, but an expression of the fact that, from the moment of inception, living matter is in a state of progressive physical and chemical change. This in turn implies a continuous flux of energy. The part that respiration plays is twofold; in the first place, it provides intermediate compounds from which further living matter may be built, and in the second place it canalises the energy flux into paths promoting the same end. It is not enough that the rich stores of energy in the foodstuffs should be 'released', as in a combustion; they must also be harnessed. These requirements, and hence the respiration that provides them, are fundamental to all protoplasm, and are side-stepped by no cells whatever.

Respiration is a spontaneous process in the sense that it proceeds with an overall loss of free energy to the organism including foodstuff. This free energy is defined as the maximal amount of energy available to do work when the foodstuff is oxidised at constant temperature and pressure to carbon dioxide and water. The adoption of the $CO_2 + H_2O$ level as the reference level for energy changes is a biological convention, and differs from the physico-chemical one. In a single spontaneously possible reaction, any free energy not used in doing work will be dissipated as heat. Muscle cells use about 20 per cent of their foodstuff free energy in doing work during contraction. But other cells rarely do work, and when they do the amount is usually negligible. Nevertheless the loss of foodstuff energy is not necessarily equalled by the production of heat. Especially during growth the free energy of the lost foodstuff molecules is transferred internally into other high energy molecules, such as the newly formed proteins, and so retained inside the cells. The relative proportions retained within the cell and lost as heat vary enormously. Retention is greatest in rapidly growing cells; but may become virtually nil in adult cells that are just 'ticking over'. Cells are not heat engines, i.e. they cannot utilise a temperature difference to do work as has been shown with considerable precision for muscle cells. Nevertheless heat release may not be useless to cells, an obvious example being the maintenance of body temperature in warm-blooded animals. At the cellular level itself a temperature maintained somewhat above the usual environmental range accelerates virtually all vital processes. Energy turnover is continuous; whether it results in retention or dissipation depends on variations in the nature and coupling of the respiratory mechanisms.

Free energy of respiratory reactions

In the commonest type of biological reaction

1) $A + B \rightleftharpoons C + D$

the free energy change (ΔG) is given by

2) $\Delta G = -RT \ln \dfrac{[C]\,[D]}{[A]\,[B]}$

where R is the gas constant, T the absolute temperature and the square brackets indicate concentrations or, more strictly, activities. The free energy change must be negative and is expressed in cals per mole. When the reaction has come to equilibrium its free energy is zero. If this only happens when the ratio of products [C] [D] to reactants [A] [B] is very high, i.e. if the reaction runs more or less to completion, the free energy under 'standard' conditions is also high. Conversely if equilibrium is reached with a low ratio the reaction runs from right to left and has a high free energy loss in that direction. The majority of biological reactions lie between these two extremes, i.e. their value of $K = [C]\,[D]/[A]\,[B]$ at equilibrium is not so far from unity and they have a small free energy in one direction or the other. Such reactions are readily reversible because small changes of conditions may change the equilibrium point and the direction of free energy loss.

Standard conditions for solutions are taken as unit concentrations (activities) of all reactants including H^+, and a temperature of $25°C$. These of course may depart markedly from normal biological states. Writing the standard free energy as ΔG_0 the free energy for other temperatures and concentrations can be calculated from

3) $\Delta G = \Delta G_0 + RT \ln [C]\,[D]/[A]\,[B]$

Since hydrogen ions are so often a reactant, the data are sometimes recalculated to give $\Delta G_0'$, for pH 7 instead of the standard pH 0. It is, however, rarely that effective H^+ concentrations can be determined at biological reaction points. Estimates of ΔG_0 have now been made for all the more important respiratory reactions and are helpful in understanding the nature of the change going on.

Activation energy

A reaction resulting in decrease of free energy does not necessarily start when the reactants are mixed. The molecules, in for example a pure solution, do not all possess the same amount of energy, and the proportion having enough energy to react may be so small that the rate of reaction is negligible. The amount of energy that must be added to bring the average molecule into the reactive state is called the energy of activation. The simplest way to provide it is to raise the temperature; but the method is not of much use to cells because they are destroyed around 40°C. The realisation that cells contain catalysts, the enzymes, that can enormously accelerate reaction rates without any rise of temperature, was the first major step towards an understanding of metabolic processes. The enzymes are true catalysts, coming out of a reaction in the same form as they enter it, i.e. they do not themselves contribute energy. They form intermediate compounds with the reactants and in so doing may be said, in a broad way, to lower activation energies. This is highly effective because a small reduction of activation energy causes a large increase in the proportion of active molecules; reducing the activation energy to a tenth may increase the number of active molecules more than a millionfold. The methods by which enzyme-substrate interactions bring about the reduction are still debateable.

It should be pointed out that the energy of activation, a property solely of the reactants, is in no way related to the free energy of the subsequent reaction which is the difference of energy level between the reactants and products. These various quantities can be diagrammatically represented as in Figure 1.

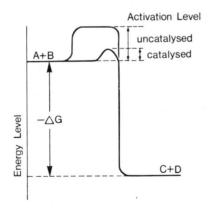

Figure 1   Diagram to illustrate energy levels in a bimolecular reaction.

The nature of respiration

The first tool available to students of respiration was measurement of its gaseous exchanges, its oxygen-uptake and CO_2-output. It is still an important tool, because it is one of the few that can be applied directly to the living cell, and may therefore be the means of deciding what is actually going on as distinct from what the cell has the equipment to do. Variations of rate with temperature, oxygen concentration and stage of development, and comparisons of CO_2-output with oxygen-uptake under all these conditions lay down a framework within which further knowledge has to be contained. But the systems involved are so complex that no more than a framework can be expected from gas kinetics alone. An enormous step forward was taken when intermediate products began to be isolated and when the enzymes catalysing the intermediate reactions were extracted and their potentialities examined. It was soon realised that many of the extracted enzymes worked much faster in solution than they could possibly be working inside the cell; and, moreover, that mixtures of enzymes did not lead in the hurly-burly of a test-tube solution to the products formed in cells. It was evident, therefore, that cell respiration was a self-regulating process in which the co-ordinating controls were of equal importance with the accelerating catalysts. At the present time we have a considerable, sometimes embarrassingly rich, knowledge gained over the past fifty or sixty years, of the catalysts and reaction paths of respiration. We have much less understanding of the controls; but the development of more sophisticated biochemical techniques and the advent of the electron microscope have made questions of organisation and control the present growing points of the subject.

By far the greatest energy turnover occurs during the final oxidation of some intermediate, usually pyruvic acid, to carbon dioxide and water. The term, respiration, is sometimes limited to this stage, and the production of the intermediate called glycolysis or fermentation. This is rather like saying that only dinner counts as eating and that breakfast and supper should be called pre- and postprandial alimentation. In the following pages we shall use respiration as an umbrella term covering the whole process that results in a net loss of free energy from foodstuffs, associated with a number of stages useful to the organism. We shall not attempt to follow through the connections of respiration with the other branches of metabolism and with growth, since to do so would involve the total of cell physiology.

The cell concept

The cell idea had a gestation period of more than two hundred years, from Hooke's first microscopic examination of cork in 1665 to Schleiden and Schwann's 'cell theory' of the 1830s. But, even then, it could only be stated that living beings consist of cells and cell products, and grow and reproduce by multiplication of cells. There was no distinct account of what a cell is and, in spite of great progress, this is still not completely possible yet.

TABLE I

Approximate mean sizes of some cells

Acetabularia mediterranea (alga)	0·2 ×	10 cm
Squid (*Loligo*) axon	0·2 ×	150 mm
Amoeba proteus	100 −	600 μm
Storage cell (apple flesh)	100 ×	100 μm
Green cell (spinach leaf)	20 ×	40 μm
Liver cell	20 ×	20 μm
Protococcus (green alga)	10 ×	20 μm
Trypanosoma sp.	4 ×	27 μm
Yeast (*Saccharomyces* spp.)	6 ×	8 μm
Anabaena (blue-green alga)	3 ×	4 μm
Red cell (human erythrocyte)	2 ×	8 μm
Bacillus	0·6 ×	2 μm
Coccus	0·2 −	1 μm

Cells come in a great variety (Table I), from the spherical bacterial coccus ($0·2–1·0$ μm) to the parasol-shaped *Acetabularia mediterranea* (1–10 cm long); and from unicellular metabolic 'all rounders' to the highly specialised tissue cells of animals and plants. For purely practical reasons of handiness and availability in bulk, some types have had their respiration studied much more intensively than most. There is no need for us to attempt any exact definition of what is a cell and what is not, and this is fortunate because in the past the effort has led to scholasticism rather than to biological progress. Instead there follows a brief description of those individuals that have played star rôles in the study of respiration. Their similarities and differences of structure, and the changes that these may undergo with changing conditions have important connections with their respiratory behaviours.

The bacterial cell

Bacteria afford the smallest and simplest of all cells. The spherical cocci are usually less than 1 μm in diameter and the rod-shaped bacilli are not more than 2 μm long. *Escherichia coli* cells are about $0·5 × 2·0$ μm with only $5–25 × 10^{-14}$g dry weight. These dimensions are very similar to those of mitochondria (p. 13) and give a volume about 1000 times less than that of most cells. In spite of this minute size, a considerable knowledge of bacterial structure has now been built up with the aid of experiment, and light and electron microscopy. We need only concern ourselves with the simpler bacilli. Many are surrounded by a capsule. This is a thin gel containing polysaccharides or polypeptides of high molecular weight, or both. The amount of capsule produced is very variable and may be nil. It is said to have a protective function and may even be reabsorbed as food. It does not readily take up dyes and is usually revealed by 'negative staining', i.e. mounting in a dark fluid such as indian ink or nigrosin.

Inside the capsule is the cell wall (Figure 2). It can be obtained in pure

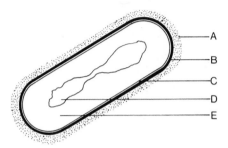

Figure 2   Diagram of bacterial cell structure. A capsule; B cell wall; C plasma membrane; D nuclear material; E cytoplasm.

preparations from which the cell contents have been removed by mechanical disintegration and differential centrifuging (Plate I*a*). It is comparatively rigid, not elastic, and is composed of proteins, polysaccharides and sometimes lipids. It is chemically inert and about 15–20 nm thick.

The wall can be specifically removed by the enzyme, lysozyme (from tears and egg white), and the protoplasts released. If the enzyme solution has been made isotonic, with sucrose for example, the protoplasts round off into spheres and remain viable at least for a time. They have an outer layer, the plasma membrane (plasmalemma) which is visible in some electron microscopic preparations (Plate I*b*). It can also be shown by plasmolysis experiments to be

---

Plate I (*a*)   Separated cell walls of *Bacillus polymyxa* × 22 000. Courtesy J. Baddiley.
(*b*)   *Bacillus anthracis*. CR residue of capsule; CW cell wall; BCM bacterial cytoplasmic membrane (plasmalemma); N nuclear material. Left × 35 000; right × 72 000. Courtesy J. E. Roth, C. W. Lewis and R. P. Williams.

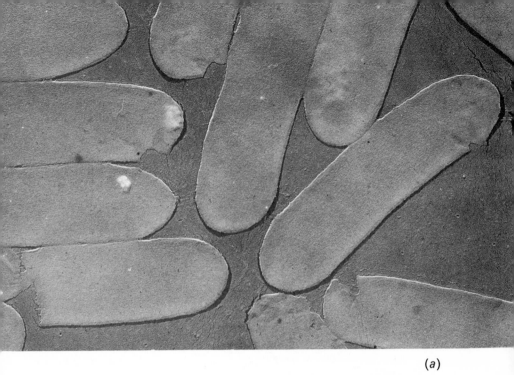

(a)

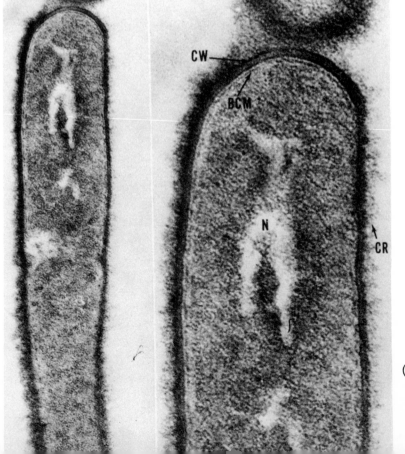

(b)

semipermeable, unlike the wall. If the isolated protoplasts are burst in hypotonic solution, the particles into which the cell membrane disintegrates can be
separated centrifugally from the internal protoplasm. They consist of 40–50
per cent protein, 20–30 per cent lipid and 15–20 per cent polysaccharide and
make up about 10 per cent of the protoplast. The membrane of *Staphylococcus
aureus* has been calculated to form a layer only 5 nm thick over the surface of
the protoplast and to be able to form one monolayer of lipid and one of protein. This thin layer carried more than 90 per cent of the recovered cytochrome of the cell and equally high proportions of many other enzymes of
respiratory oxidations (Table 2). In the unusually large cells of *Thiovolum
majus* the plasma membrane has infoldings deep into the cytoplasm. When we
come to the higher cells, it will be seen that the bacterial cell membrane has
much in common with the inner membranes of their mitochondria (p. 13).

TABLE 2

Percentage distribution of some enzyme activities in *Staphylococcus aureus* (data of
P. Mitchell).

Enzyme	Plasma membrane (lipoprotein)	Protoplasm
Cytochromes	90	10
Lactic dehydrogenase	80–95	5–20
Malic dehydrogenase	90	10
Malic enzyme	90	10
Formic dehydrogenase	90	10
α-Glycerophosphate dehydrogenase	50–70	30–50
Glucose-6-phosphate dehydrogenase	3	97
Glucose-6-phosphatase	10	90
Acid phosphatase	90	10

The cytoplasm lying inside the plasma membrane is seen in electronmicrograms stained with osmium to contain numerous more or less evenly
distributed particles of 10 to 20 nm diameter. They are believed to contain
ribonucleic acid (RNA). Readily stainable 'volutin' granules and lipid droplets, may also be present. If there are flagella, there are relatively large 'basal
granules' below their points of attachment. It is claimed that the reduction of
tetrazolium appears to be localised and may be associated with dense bodies
in, for example *E. coli*; but these bodies have only been seen after active
growth is past, which does not suggest a basic respiratory function for them.
　　DNA does not stain with osmium and the usual electron pictures show a
clear space of irregular outline in the centre of the cell (Plate I*b*). This
'vacuole' reacts strongly with the Feulgen DNA reagent and must consist

largely of DNA without protein or histone. Under favourable conditions it can be seen to be finely filamentous and may indeed consist of a single tangled DNA molecule with an estimated length in *E. coli* of about 1 mm (i.e. about 500 times the length of the entire cell) and a molecular weight around 2000 million. There is as yet no evidence that this naked DNA is separated from the cytoplasm by any membrane.

Blue-green algae

The cells of blue-green algae (Cyanophyceae), whether free or filamentous, are a good deal larger than the bacteria (Table 1), but they have many structural features in common. In many important respects they resemble the bacteria more than they do the simpler green algae for example. They may have a slime layer, like many bacteria, and an envelope inside this with a complex structure of several layers. Within this again there is a plasma membrane.

The cytoplasmic matrix itself shows two zones (Plate II). The outer zone contains numerous lamellae which carry the photosynthetic pigments, and probably some of the 'insoluble' respiratory enzymes also. They may have the general structure of the unit type described in the next section and tend to lie in pairs forming rambling patterns. The central region has none of these lamellae; but it probably contains all the nuclear material of the cell. There are regions of low electron density which strongly resemble the DNA-containing regions of the bacterial cell. Both inner and outer zones contain numerous ribosomes about 15 nm in diameter. There are no membrane-bounded chloroplasts or mitochondria, no endoplasmic reticulum and no membrane separating the nucleoid region from the rest.

Cell membranes

Procaryotic cells, such as those of bacteria, appear to be of comparatively simple structure: and no complex system of internal membranes has yet been discovered in them. This may be associated with their minute size (cf., Table 1). The eucaryotic cells now to be described are characterised, on the other hand, by an astonishing proliferation of membrane structures. Not only is the nucleus separated from the cytoplasm by a membranous envelope, but plastids and other cytoplasmic organelles are similarly bounded. In addition, the cytoplasm is penetrated in all directions by an endoplasmic reticulum of membranes which may be deviously continuous from the cell surface to the nucleus.

The idea that the cell surface might be rich in lipids is an old one. It was invoked by Overton in 1895 to account for the normally high resistance to the

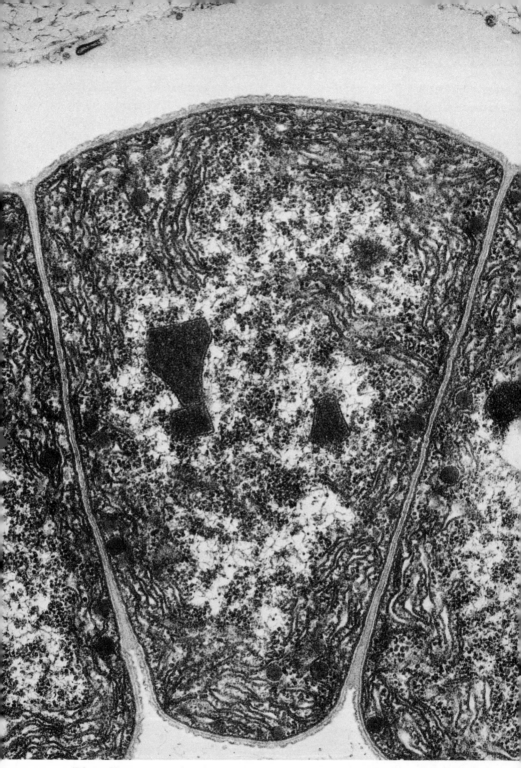

Plate II A cell of the blue-green alga *Nostoc*. The outer cytoplasmic zone has numerous paired lamellae. The inner 'nuclear vacuole' shows fine threads of DNA. Ribosomes are numerous in both zones. × 56 500. Courtesy H. B. Griffiths.

entry of water-soluble substances and the easy entrance of fat-soluble ones from outside. Nevertheless, surface tensions at cell boundaries were found to be more akin to those of proteins, and in 1943 Davson and Danielli proposed a model consisting of a bimolecular layer of polar lipids, with their hydrophilic groups adsorbing monolayers of extended protein on either side. As such a membrane would be virtually impermeable to solutes, 'aqueous pores', lined with polypeptides, were provided (Figure 3). Further proteins, either

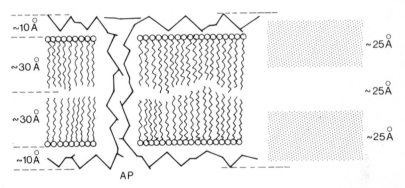

Figure 3 The 'unit membrane'. On the left a diagram of the Davson-Danielli model; AP aqueous pore. Small circles represent hydrophile ends of lipid molecules; 'tails' the lipophile ends. The heavy zig-zag represents extended protein. On the right the profile seen in electron micrographs after permanganate treatment.

stretched or globular, might be adsorbed on either surface. Confirmation of this basic pattern has been obtained from X-ray diffraction studies of myelin sheath, red cell membranes and artificial systems made with the extracted lipids. The dimensions, as indicated in Figure 3, bring the structure above the limits of resolution of the electron microscope; but at first there was some difficulty in reconciling the patterns observed in ultra-thin sections with the model. If, however, permanganate is used as the fixative, surface membrane are consistently delineated in profile as a triple structure of two dark lines each about 25 Å thick separated by a pale zone of similar width. J. D. Robertson has suggested that the permanganate is bound along the regions of attachment between protein and lipid so that the pale zone represents the non-polar regions of the lipid molecules. He calls the whole array the unit membrane and its dimensions are in good agreement with the Davson–Danielli model (Figure 3). This unit membrane structure is shown by many internal membranes as well as cell surfaces and appears to be of very widespread occurrence, especially where the main function of the membrane is semipermeability or electrical insulation.

It is, however, known that many enzymes are membrane-bound and some with great tenacity. It seems clear that further assumptions have to be made in such cases. By means of the freeze-etch technique, sections can be obtained in the plane of the membrane surface. Whether the break occurs at an outer surface of the membrane or in the internal lipid layer—at present a debatable point—the Davson–Danielli model would predict a uniform finely granular electronmicrogram. In fact, an irregular scattering of pustules above, or below, the general surface is always observed. The size and distribution of the pustules varies with the material, but they are found in synthetic lipid-protein lamellae, as well as in natural membranes.

It has been found that preparations of lecithin and cholesterol in ionised media form rings and tubes, about 110 Å across, consisting of spirals of lipid particles with a diameter of about 40 Å. Similar microtubules have been resolved in cytoplasm and are too tight to be formed of a wrapped-round Davson–Danielli membrane. Arrays of particles of similar size have been found in the surfaces of some cells, such as intestinal epithelial cells, and it is possible that the freeze-etch irregularities are due to such areas. The profile could be diagrammatically represented as in Figure 4. The micellar areas (Figure 4A)

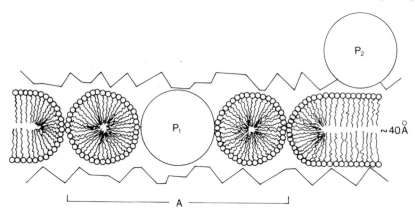

Figure 4   Diagram illustrating a particulate area of a membrane. A porous particulate area; P_1 globular protein within the membrane; P_2 globular protein attached to one side.

would have a high permeability and it has been calculated that they could not exceed 1 per cent of the surface of a red blood cell. It seems likely, moreover, that changes could occur locally from the lamellar to the micellar state and *vice versa*.

Enzymes in the form of globular proteins may be associated with such membranes in at least two ways. Small protein molecules of MW 10 000–20 000 have dimensions similar to the 40 Å diameter observed for the lipid micelles; e.g. cytochrome-c monomer (MW 12 000) is 39 × 28 Å, and myoglobin (MW c. 17 000) is a disc 45 × 35 × 25 Å. Such molecules could be fitted into the micellar membrane with very little distortion of the pattern (cf. Figure 4 P_1).

Alternatively, globular proteins could be attached on either side of the lipid membrane (P_2). In the first position they might catalyse transport reactions across the membrane and in the second position reactions localised to whichever side of the membrane they were on.

Mitochondria

The most important membrane systems at present known to be associated with respiration are the mitochondria. They can be seen in living cells treated with Janus Green B under the light microscope, though uncomfortably near to the limits of its resolution. They appear as minute dots (0·5–1·0 μm in diameter) or as fine threads about ten times as long. They may show movements and frequent changes of shape and appear able to unite end to end. When extracted and fixed they are usually more or less spherical; but within the cell their form may be quite irregular, or spiral (in sperm cells) or slab shaped (in some flight muscles) to fit the cell conformation.

Their structure has been studied chiefly in ultra-thin sections stained with osmium or permanganate, and is revealed as having two separate membranes, one enclosed within the other. Permanganate fixation shows the profile of Figure 5, i.e. of two unit membranes separated by a space about 100 Å wide.

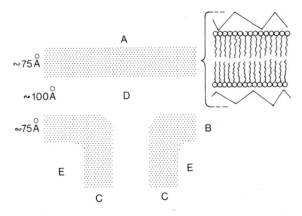

Figure 5 Diagram of the mitochondrial wall at the point of attachment of a crista. A the outer membrane; B inner membrane; C walls of crista; D aqueous space between the two membranes; E the mitochondrial matrix. On the right the outer layer is shown expanded as a unit membrane; the inner layer carries numerous enzymes and is probably much modified.

The inner membrane has many infoldings deep into the centre. In liver mitochondria, the first to be described in detail, serial sections showed these infoldings to be laminar and they were called cristae. In other mitochondria, the attachment of the cristae to the inner wall may not run all along their length;

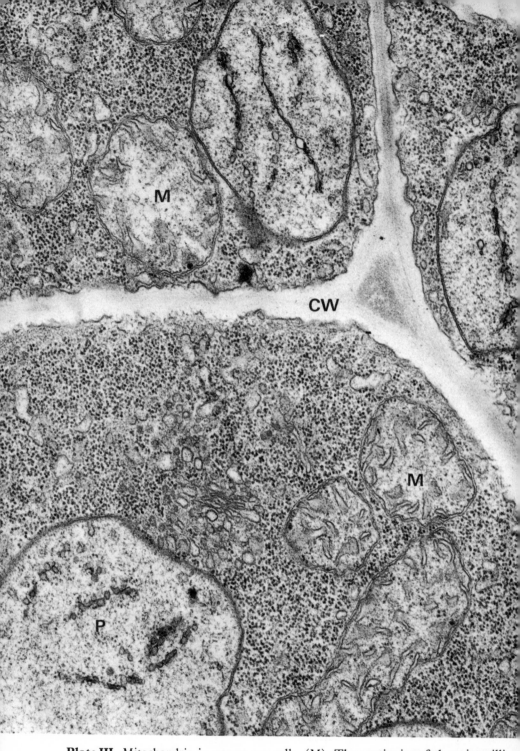

Plate III Mitochondria in young pea cells, (M). The continuity of the microvilli with the inner mitochondrial wall is visible in several places. CW cell wall. Ribosomes, the clusters of black dots, are large and numerous in the cytoplasm, smaller and less numerous in the mitochondria and proplastids, (P). × 51 000. Courtesy A. D. Greenwood.

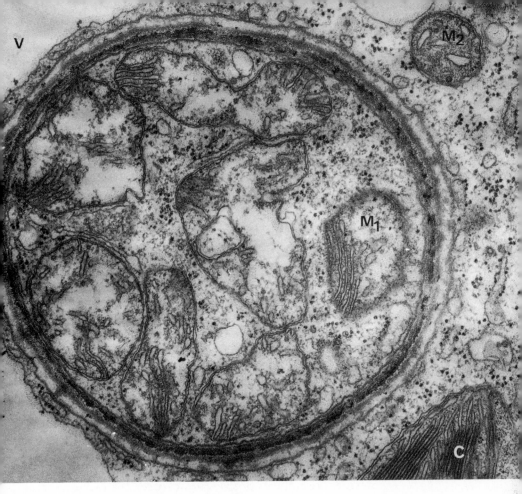

Plate IV T. S. haustorium of rust fungus *Uromyces appendiculatus* in cytoplasm of green leaf cell of *Phaseolus vulgaris*. M_1 mitochondria in the fungus with plate-shaped cristae; M_2 mitochondria in leaf cell with tubular microvilli; C chloroplast; V vacuole of leaf cell; T tonoplast of leaf cell. The dark dots are ribosomes. $\times$ 38 000. Courtesy N. V. Hardwick and A. D. Greenwood.

but be limited to a small part of it. In others the infoldings are tubular and they are then called microvilli (Plate III). The form and abundance of these infoldings show immense variation with species, tissue, stage of cell development and treatment (cf., Plates III, IV, IX and XII). In general, they are most numerous and extensive in cells with high oxidation rates and are believed to carry the aerobic oxidation systems. It has been calculated that about 25 per cent of the proteins of mitochondrial membranes are enzymes of oxidation. About 55 per cent is described as structural protein, extremely insoluble at pH 7 and with monomers of MW around 20 000–30 000. These proteins form complexes with many enzymes including cytochromes *b*, *c*, and *a*; but not directly with cytochrome *c*. They also unite firmly with phospholipids at pH 7, but maintain the membrane structure, at least in outline, after removal of the phospholipids with aqueous acetone. All the membrane proteins, both structural and enzymatic, even when they are broken loose with detergents, rapidly form insoluble polymers, a fact which has greatly increased the difficulties of studying their reactions.

The fine structure of the inner membrane and its extensions is obviously of the greatest interest. In thin sections it does not differ conspicuously from that of the outer membrane (cf. Plates III, IV, IX, X(b) and XII(a)). If it really is the seat of the oxidising systems, it seems unlikely to have the structure of a unit membrane as represented in Figure 3, and some elaboration of the type suggested in Figure 4 appears more probable. This is supported by the appearance of the cristae in freeze-etch preparations as shown in Plate XII(b) (p. 31). Particulate areas figure prominently. In isolated mitochondria negatively stained, i.e. allowed to dry out in the presence of an electron-dense medium, the deformed cristae appear to have pale edges to which numerous spherical particles are attached by short stalks. When mitochondrial membranes are disrupted, they tend to break down to particles of about the same size as these spheres. It is difficult at present to reconcile the results of the different techniques and to decide which is producing a misleading artefact and which the most significant appearance. The biochemical approach to the problem is described on p. 124.

Little is known of the contents of the space between the two membranes; but the central cavity has a matrix containing salts and soluble proteins including enzymes. DNA has also been identified in the mitochondrial matrix of many plant and animal cells. The fibrils may be clumped or dispersed and a fibril has been obtained from the mould *Neurospora crassa* with a length of 6·6 μm and MW about 13×10^6. Although lying mainly in the matrix, mitochondrial DNA may have occasional attachments to the inner wall, which also contains small amounts of a characteristic RNA.

From a respiratory point of view the main function of mitochondria is undoubtedly the oxidative formation and export of ATP; but it would be misleading to think of this as their only activity. They are able to synthesise proteins. They also show energy-coupled swelling and contraction, and their walls contain an actomyosin akin to the protein of muscle fibrils. They accumulate ions by active transport mechanisms and so may act as regulators of concentrations in the general cytoplasm. These various activities are so closely integrated in mitochondrial membranes, that no separations of them have yet been achieved. The respiratory aspects will be considered more fully in Chapters 5 and 9.

Trypanosomes

The *Trypanosoma* spp. are a group of Protozoa that have been intensively studied because some of them are important pathogens in men and animals including game. There are other species which live harmlessly in fishes and amphibians as well as in various land vertebrates. They all require alternative hosts, those infesting fishes and amphibians are usually carried by leeches, the others by insects. *T. rhodesiense* and *T. gambiense*, the causes of African sleeping sickness, exist and multiply in the alimentary tract and later in the salivary glands of the tse-tse fly and may be introduced into the bloodstream of men, or some stock animals, when the fly is sucking. They thus have two phases, with a temperature difference of about $10°C$. When introduced into the warm blood stream they become very active indeed, with a respiration rate perhaps fifty or more times that of the host (Table 3).

TABLE 3

Respiration Rates
$\mu l\ O_2$ (or CO_2)/h mg dry weight

Cells and Tissues	
Trypanosoma lewisi	68
Yeast (aerobic)	60–100
Yeast (anaerobic)	100–300
Arum italicum spadix slices	31
Kidney slices	21
Liver slices	9
Barley, 7 day seedlings	1·6
Ascaris suis muscle pulp	1·3
Peas (air dry)	$1·2 \times 10^{-4}$
Isolated organelles	
Mitochondria, housefly muscle	400–600
Mitochondria, liver	40–80
Nuclei, calf thymus	< 0·5

Trypanosoma cells are 20 times longer than the average bacillus and have a more obviously complex structure. They are long and narrow and their length is further prolonged by a flagellum which is attached along the edge of an undulating membrane until it enters the main body of the cell at the hinder end (Figure 6). The flagellum contains the usual 11 longitudinal fibrils arranged as a central pair with 9 surrounding each of about 37 nm diameter. Close to the basal granule of the flagellum there is a 'kinetoplast', surrounded by a membrane and containing a number of electron-dense lamellae (Plate V*a*). The most conspicuous organelle is the large nucleus, which, in contrast to the DNA-

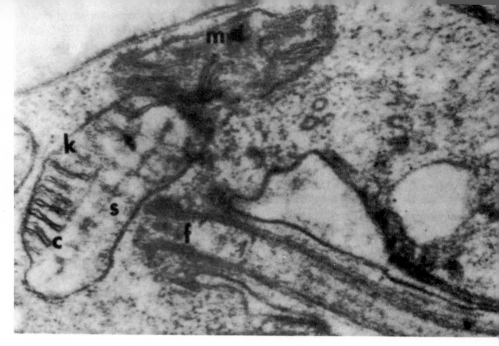

Plate V (*a*) *Trypanosoma lewisi*. Base of flagellum (f) and kinetoplast (k). m, mitochondrion; c, cristae; s, 'folded membrane'. × 50 000. Courtesy D. M. Judge and M. S. Anderson.

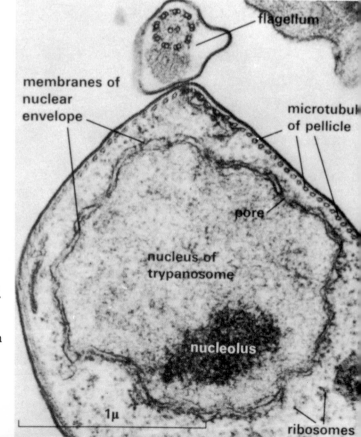

(*b*) *Trypanosoma brucei*. Region of the nucleus, × 50 000. Courtesy K. Vickerman and F. E. Cox.

containing area in a bacillus, is separated from the cytoplasm by a well defined membrane (Plate V*b*). The nucleus appears in electronmicrograms as a finely granular coagulum, suggesting the presence of proteins, with an aggregation of denser particles, the nucleolus. The general cytoplasm appears finely granular with an irregular scattering of 'volutin' granules, some at least of which may be ribosomes including ribonucleoproteins, since they are re-removed by ribonuclease. The whole cell is enclosed in an outer membrane which sometimes appears double and which has fine spiral striations due to microtubules about 15 nm thick and spaced about 10 nm apart.

When the trypanosomes are first introduced into a warm blood stream they have the 'slender' form shown in Figure 6, and they continue in this state so long as they are rapidly dividing by a binary fission. Eventually division slows

Figure 6 *Trypanosoma brucei* as seen under the light microscope in the 'slender' stage Length about 30 μm. (After D. Bruce.)

down and the form becomes 'stumpy'. Mitochondria are probably present at all stages though, as will be seen later, respiration during the actively dividing period is almost wholly anaerobic. Correlated with the changes of form and activity there are associated changes in the mitochondria. *T. rhodesiense* is said to have a single mitochondrion which runs the full length of the cell and makes close contact with, or perhaps even includes, the kinetoplast at the base of the flagellum. (cf., *T. lewisi* in Plate V*a*). The mitochondrion has the usual two membranes, but during the cell's active long, slender stage, it is said to lack internal cristae. Later when activity is slowing down and the cell is becoming stumpy, the mitochondrion develops cristae with, presumably, the usual respiratory enzymes. It is conjectured that this occurs when for one reason or another the sugar supply from the host becomes more restricted and a purely anaerobic respiration is inadequate for the survival of the cell. Similarly, under the less favourable conditions represented by the gut of the fly, the mitochondrion is described as being an extensive network of canals radiating out from the kinetoplast and containing numerous cristae.

Yeasts

The yeasts that have contributed so much to the study of cell respiration are all strains of *Saccharomyces cerevisiae*. They include the fully aerobic bakers'

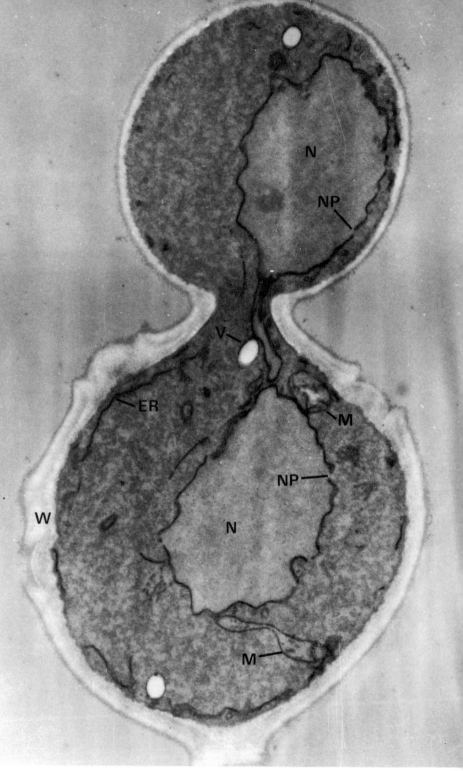

Plate VI Yeast cell, budding. Thin section showing: N nucleus; NP nuclear pores; M mitochondria; V vacuoles; ER endoplasmic reticulum; W wall. × 21 800. Courtesy J. Gay.

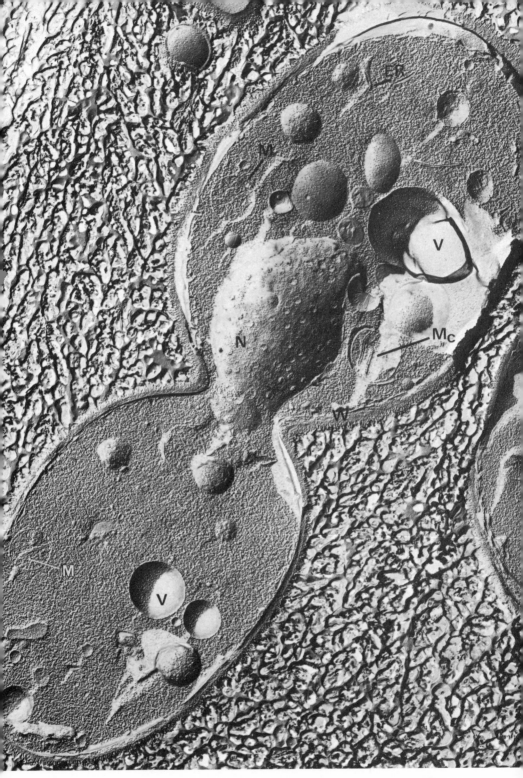

Plate VII Yeast cell, budding. Freeze-etch preparation showing: N nuclear surface with pores; M mitochondria; M_c mitochondrion with conspicuous cristae; V vacuoles; ER endoplasmic reticulum; W wall. × 19 500. Courtesy H. Moor.

yeasts and the more anaerobic brewers' yeasts. Since the quality of beer depends very markedly on the particular yeast employed, many pure strains
have been isolated and maintained with great care. They fall into two classes;
the quickly respiring top yeasts giving beers of high alcohol content and the
more slowly respiring bottom yeasts giving lagers.

 S. cerevisiae is unicellular probably by reduction from a filamentous ances-
tor. The cells are ellipsoidal, about 8×10 μm, and multiply by budding
(Plates VI and VII). They are surrounded by a definite wall, at first thin and
elastic, but becoming thicker as the cell ages. The mature wall probably has
two layers and consists mainly of glucose and mannose anhydrides with some
proteins, fats and mineral substances. Sometimes there is a poorly defined
outer layer recalling the capsule of bacteria. The cytoplasm has a fine plasmalemma in contact with the wall. The wall itself can be removed experimentally by cautious treatment with snail juice and the cytoplasmic surface
then remains semipermeable. When appropriately stained, the cytoplasm is
seen to be crammed with ribosomes, particles of about 15–25 nm diameter
and rich in RNA. The cytoplasm may also contain a number of glycogen
granules. There is usually a large vacuole, often containing particles in
Brownian movement some of which may be lipid droplets. Adjacent to the
vacuole is the nucleus, which is filled with fine-grained chromatin with a
crescent of scattered denser particles. At present any arrangement of the
chromatin into chromosomes remains debateable. The nucleus is separated
from the cytoplasm by an envelope of two membranes (Plate VII), penetrated
by about 200 pores estimated to occupy around 6–8 per cent of the surface.
The pores may be occupied by fine-grained matter apparently different from
that of either nucleus or cytoplasm. There is an endoplasmic reticulum which
is continuous with the nuclear membrane and which is often particularly
conspicuous below and parallel with the plasmalemma (Plate VI). Connections between this endoplasmic reticulum and the plasmalemma have not,
however, been observed. There are usually also mitochondria in the cytoplasm (Plates VI and VII); but their number, size and structure vary
greatly with cultural conditions and in parallel with changes of respiration.
These will be considered in the appropriate places.

Plant cells

Young, rapidly respiring, plant cells (Q_{CO_2} 5–10) are characterised by a thin
cell wall and a large nucleus (Plate VIII*b*). There is a central vacuole, which
increases very rapidly as the cell enlarges and which is separated from the
cytoplasm by a semipermeable division which appears to be a single unit
membrane (p. 11). The vacuole contains proteins, organic acids, salts and

other solutes. The cytoplasm becomes a thin layer (1–5 μm), pressed between the wall and the vacuole, and has a plasma membrane against the wall, with occasional infoldings into the protoplasm. The wall, at such a stage, may be almost pure cellulose, though it is laid down on a protein and pectin middle lamella. The separation of adjacent cells is much less complete than might appear as fibrils of protoplasm (plasmodesmata) connect them through the walls at frequent intervals, and the cellulose gel of the wall itself is about as permeable to solutes and gases as an equivalent layer of water.

The large nucleus contains numerous coiled chromosome threads. In the early stages of interphase there is relatively little DNA, the amount of which builds up as mitosis approaches. In considering the relation between the nucleus and respiration it is important to remember that interphase is the period when the nucleus is growing, i.e. synthesising proteins, nucleotides, etc. Extracted wheat nuclei can about double their DNA content, but not more. The nuclei contain one or more nucleoli, dense bodies which are not separated from the nucleoplasm by any membrane (Plate VIIIa). They are so much denser than the nucleoplasm that they can be centrifuged out of it. They are rich in protein and RNA; but appear to contain little or no DNA. The nuclei are separated from the cytoplasm by an envelope, similar to that in yeast, and with similar wide pores, 20–50 nm across. Its elasticity is so high that it can expand and contract twofold. It does not hold the nucleus together. Extracted nuclei that have largely lost their envelopes (Plate VIIIa) remain coherent and capable of syntheses. The outer membrane of the envelope is also directly linked with the endoplasmic reticulum (ER) which penetrates the cytoplasm in all directions.

The cytoplasm probably has as matrix a thin colloidal gel in which are immersed numerous organised membranes, either ramifying like the endoplasmic reticulum, or delimited in special organelles such as the mitochondria.

A young plant cell will also contain proplastids, of size similar to that of mitochondria (0·5–5 μm), all surrounded by membranes; but with internal structures already differing according to the direction of their future development. Not all specialised regions are separated from the general cytoplasm by enclosing membranes; a notable exception being the dictyosomes (Golgi bodies). These are areas of smooth membranes, without adhering ribosomes, enclosing small empty-looking vesicles. Their appearance is very similar in both plant and animal cells (Plate XI) and they have no membrane separating them from the 'ground' cytoplasm. In young plant cells they are especially noticeable around the periphery when cell division is just complete and the new cell plate is forming across the middle of the spindle.

It is probable that the full extent of cytoplasmic differentiation is not yet known. New organelles are still being discovered, of which two from plant

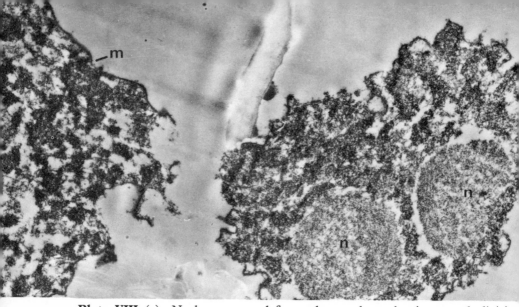

Plate VIII (*a*) Nucleus extracted from wheat embryo showing 2 nucleoli (n), chromatic material and ruptured membrane (m). × 17 700. Courtesy A. D. Greenwood.

(*b*) Young cells from pea seedling. N nucleus; n nucleolus; V vacuole; M mitochondria; P proplastids. × 3000. Courtesy A. D. Greenwood.

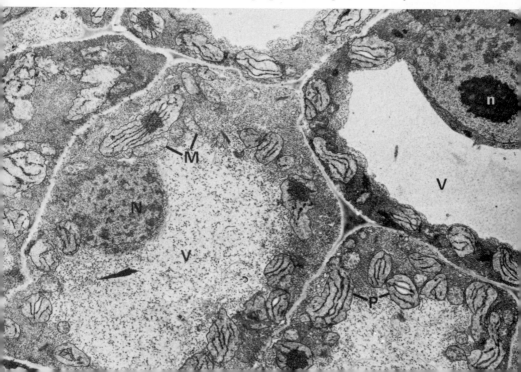

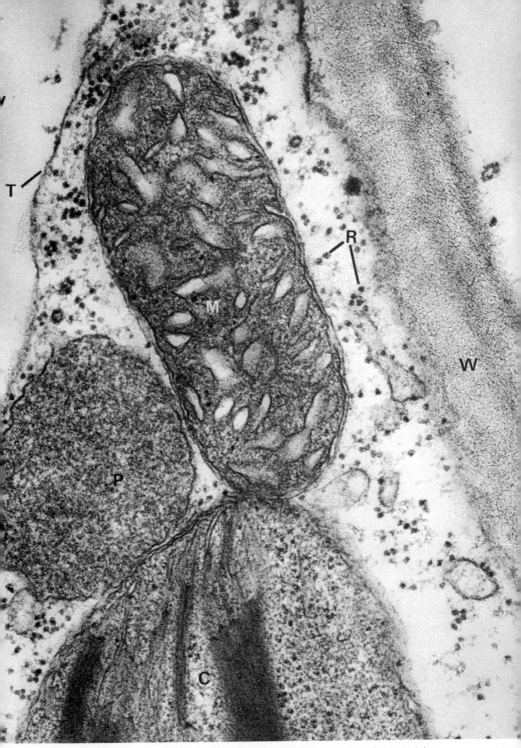

Plate IX Part of green cell of *Phaseolus vulgaris*. C chloroplast; M mitochondrion; R ribosomes in cytoplasm, smaller ones are visible in both chloroplast and mitochondrion; P microbody (peroxisome?); V vacuole; T membrane separating vacuole and cytoplasm. W cell wall. × 52 600. Courtesy A. D. Greenwood and N. V. Hardwick.

cells may be mentioned. The peroxisomes (glyoxysomes) (Plate IX) are similar in size to mitochondria, but lack their internal structure. Their stroma has a finely granular appearance in electronmicrograms and they are bounded by a single unit membrane. They are characterised biochemically by a high content of glycollic oxidase and catalase. The sphaerosomes isolated from plant cells, also lack internal structure, but are characterised by a high lipid content and have been regarded as 'lipid-buffers' for the protoplasm, since their size varies inversely with its activity.

Ribosomes are abundant in the cytoplasm of young plant cells (Plate III). They appear to occur freely in the soluble cytoplasm, or as polysomes, i.e. aggregates of various sizes, or attached to the endoplasmic reticulum to give the so-called 'rough membrane'. They consist of about half protein and half RNA and are destroyed by ribonuclease. They are generally regarded as the sites of some of the protein syntheses.

The endoplasmic reticulum of membranes is probably universal in plant cells; but becomes less obvious as the cells mature and is much more conspicuous in many animal cells (p. 28).

Storage cells

Storage tissues, such as potatoes, carrots and apples have often been used in respiration studies, because they provide a large bulk of relatively uniform cells. They have very large vacuoles or plastids, containing reserve sugars or starch, and their respiration rates are low (Q_{CO_2} around 1). The cytoplasm is a very thin layer lining the cell wall, but is known to contain nuclei, mitochondria and other organelles.

The green cell

The green cells of higher plants possess all the usual organelles already mentioned, but their proplastids are fully developed as chloroplasts (Plates VIII*b* and IX). These contain the membranes, on which the photosynthetic pigments are located, and which are embedded in a stroma of finely granular structure. The chloroplast is surrounded by an envelope consisting of at least two unit membranes. It does not appear to have any pores, linking the stroma with the soluble cytoplasm, comparable with the pores in the nuclear envelope. Chloroplasts can be extracted with their envelopes more or less intact and can then be caused to photosynthesise at rates comparable with that in the cell. They are normally included in the cytoplasm and are pressed with the rest of it against the cell wall. The mitochondria, which are smaller, occupy positions between the chloroplasts and are inevitably often pressed closely against them.

The photosynthetic apparatus of the higher green cell, unlike that of the protocells (photosynthetic bacteria and blue-green algae), is firmly delimited from the general cytoplasm by the lipoprotein membranes of the envelope. Since the chloroplasts burst in hypotonic solutions, these membranes are clearly semipermeable; but the extent to which they hold photosynthesis and respiration apart in a green cell is still hard to assess.

Animal cells

The tissue cells of animals do not have a cell wall and do not develop large central vacuoles. In a not too specialised example, such as a liver parenchymal cell (Figure 7) there is a large nucleus separated from the cytoplasm by the usual double envelope, the outer membrane of which is continuous with the endoplasmic reticulum. The clear space between the two membranes may be continuous with the cisternae and vesicles of the reticulum which is some-times regarded as forming an intracellular transport system. The outer sur-

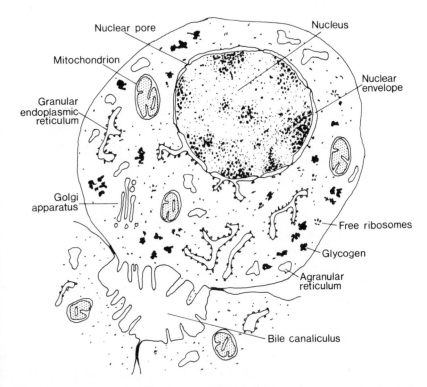

Figure 7 Diagram of frog liver cell showing structures revealed by electron micro-scopy. (After E. G. Gray.)

faces of the ER membranes, towards the general cytoplasm, are sometimes rough, i.e. studded with numerous ribosomes (Plate X*b*). In cells secreting protein, rough membrane becomes much more conspicuous and may be closely packed. Smooth membrane, without ribosomes, is said to be most plentiful in cells segregating products other than proteins. Both forms occur in liver cells. Golgi bodies of typical form (Plate XI) are also found. There are also numerous mitochondria, and also lysosomes of similar size, but without the same internal membrane structure and characterised biochemically by a high content of acid phosphatase, ribonuclease and other hydrolases. Liver cells store glycogen and this appears as small granules scattered through the cytoplasm. There may also be lipid droplets.

Animal cells are bounded by a plasmalemma, probably a single unit membrane. Adjacent cells of a tissue may be separated by a space, about 10–20 nm across, or they may be so closely appressed that the outer surfaces of their membranes appear to be fused. (Plate X*a*). There do not appear to be cytoplasmic continuities from one cell to the next as in plant cells; but where rapid transmission from one cell to another is important the two surfaces may be thrown into interlocking folds. Where cells abut upon a fluid in a canal the surface may be increased by numerous papillae (Plate X*b*).

Muscle cells

The cells of striated muscle, as in heart, limb or breast muscles, are the most highly specialised that we have to consider. The cells unite into long fibres, usually cylindrical with tapering ends. They may be united into a syncytium containing numerous nuclei, without separating cell surfaces; but the electron microscope has revealed at least some separations. The tapering ends of the long cylindrical cells interlock at irregular intervals. The centre of the cell is occupied by a mass of contractile and sliding fibrils running in the direction of the long axis of the cell (Plate XII*a*). The cytoplasm (sarcoplasm) runs in irregular patterns over the surface of the fibrillar mass. In heart muscle, the sarcoplasm is almost wholly occupied by the numerous large mitochondria with unusually abundant cristae (Plate XII*b*), and is a favourite source for extraction of respiratory enzymes. At special positions it sends processes containing endoplasmic reticulum into the fibril mass. Surface membranes can be seen where the cells abut on capillaries or connective tissue.

Respiration is not linked to a single organelle in the way that photosynthesis is limited to the chloroplast. Even in the more specialised types it is a function of the cell as a whole. The foregoing sketch has aimed to show the sort of structures among which it is going on; the more detailed connections will be developed as we proceed.

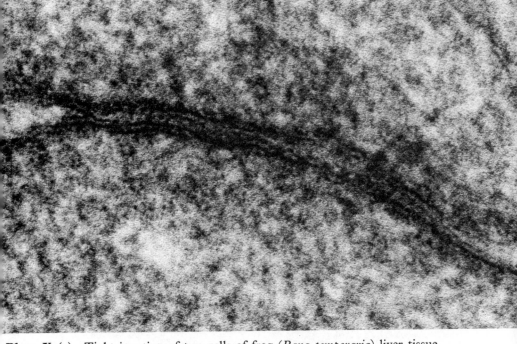

Plate X (*a*) Tight junction of two cells of frog (*Rana temporaria*) liver tissue, showing unit membranes of plasmalemma. × 375 000. Courtesy R. A. Willis.

(*b*) Bile canaliculus (BC) from frog liver, showing intrusions of cell surfaces. T tight cell junction; M mitochondrion; RE 'rough' endoplasmic reticulum with attached ribosomes. × 41 000. Courtesy R. A. Willis.

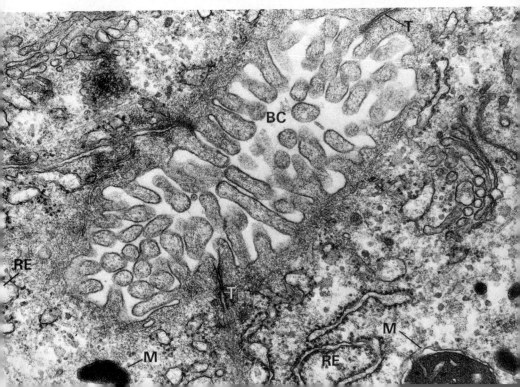

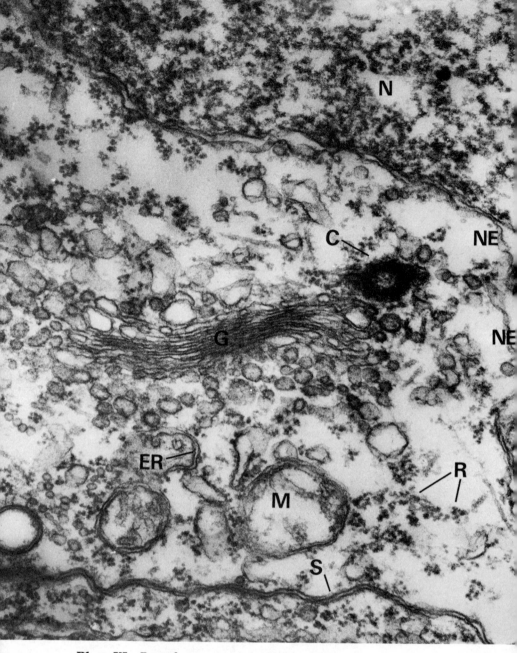

Plate XI Part of octopus nerve cell. M mitochondrion; N nucleus; NE nuclear envelope showing space between the two membranes; C centriole; G Golgi apparatus; ER endoplasmic reticulum; R free ribosomes; S surface membrane. × 45 000. Courtesy E. G. Gray.

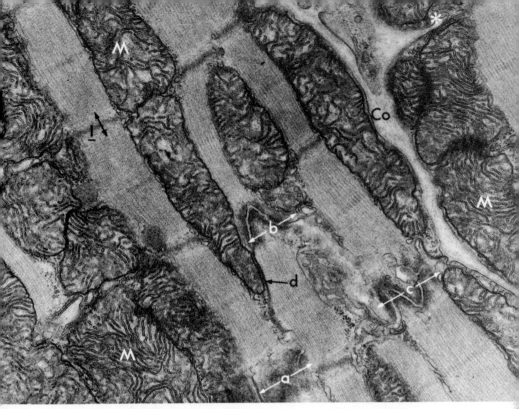

Plate XII (*a*)   Section of heart muscle of bat.
a, b, c intercalated disc joining two cells; d plasmamembrane; I contracted I band; M mitochondria; Co connective tissue. × 26 000. Courtesy K. R. Porter and M. A. Bonneville.

(*b*)   Isolated mitochondria from heart muscle of mouse. Freeze-etch technique showing cristae in cross section. × 64 000. Courtesy H. Moor.

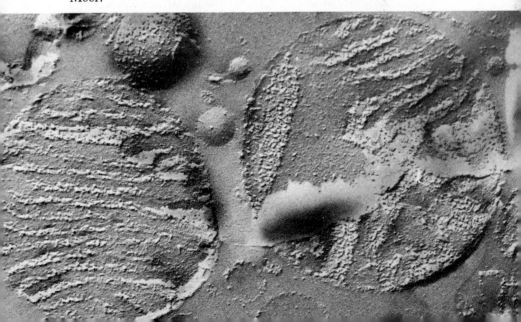

Further Reading

Anderson, E., Saxe, L. H. and Beams, H. W. (1956) Electron microscopic observations of *Trypanosoma equiperdum. J. Parasitol.*, **42**, 11–16.

Branton, D. (1969) Membrane Structure. *Ann. Rev. Plant Physiol.*, **20**, 209–38.

Burton, K., and Krebs, H. A. (1953) The free energy changes associated with the individual steps of the tricarboxylic acid cycle, glycolysis and alcholic fermentation and with the hydrolysis of the pyrophosphate groups of adenosinetriphosphate. *Biochem. J.*, **54**, 94–107.

Clowes, F. A. L. and Juniper, B. E. (1969) *Plant Cells*. Blackwell Scientific Publications, Oxford.

Fogg, G. E. (1968) *Photosynthesis*. The English Universities Press, London.

Glauert, A. M. (1968) Electron microscopy of lipids and membranes. *J. Roy. Microscop. Soc.*, **88**, 49–70.

Goldsby, R. A. (1967) *Cells and Energy*. Macmillan, London.

Gray, E. G. (1964) Electron microscopy of the cell surface. *Endeavour*, **23**, 61–5.

Jensen, W. A. and Park, R. B. (1967) *Cell Ultrastructure*. Wadsworth, Belmont, Calif.

Klotz, I. M. (1967) *Energy Changes in Biochemical Reactions*. Academic Press, New York.

Lehninger, A. L. (1964) *The Mitochondrion*. Benjamin, New York.

Loewy, A. G. and Siekevitz, P. (1970) *Cell Structure and Function*. 2nd ed. Holt, Rinehart and Winston, London.

Marchant, R., and Smith, D. G. (1968) Membranous structures in yeast. *Biol. Rev.*, **43**, 459–80.

Picken, L. E. R. (1960) *The Organisation of Cells and other Organisms*. University Press, Oxford.

Porter, K. R. and Bonneville, M. A. (1969) *An Introduction to the Fine Structure of Cells and Tissues*. 3rd ed. Henry Kimpton, London.

Roodyn, D. B. (1967) The Mitochondrion; in *Enzyme Cytology*. Academic Press, New York.

Wildon, D. C. and Mercer, F. V. (1963) The ultrastructure of the vegetable cell of blue-green algae. *Aust. J. Biol. Sci.*, **16**, 585–96.

2 *Hard-Core Respiration*

RESPIRATION shows many variations of mechanism with time and place; but one of its most striking features is that it presents a central series of reactions that occurs in almost all cells from the simplest to the most advanced and at all stages of their development. This series is independent of oxygen, and is found in aerobic and anaerobic organisms alike. It enables some normally aerobic organisms with a relatively low energy requirement, such as the higher plants, to survive considerable periods without oxygen; but not those so closely adapted to aerobic conditions as the higher animals.

The most pampered organisms are those that live in the blood stream of another. Readily assimilable and respirable foodstuffs are presented to them at an optimal temperature, and undesirable waste products removed with equal efficiency. They need only minimal transfers of energy and can get by with the minimum of metabolic equipment. A pointed example is afforded by some of the trypanosomes, which when they get into the human blood stream cause sleeping sickness. *Trypanosoma gambiense* and other species of the *brucei* group absorb glucose from the blood and convert it almost quantitatively to pyruvic acid, which in spite of the presence of the oxygen in the blood is not further broken down.

Other species such as *T. rhodesiense* form pyruvic acid with an admixture of glycerol. So long as glucose is supplied, the cells remain mobile and on transfer to another host cause a 'fulminating infection', i.e. they multiply rapidly. If the glucose supply is stopped they cease respiring, lose mobility and cell disintegration sets in rapidly. In short, the series of reactions from glucose to pyruvic acid provides all these organisms need in the way of respiration.

The full details of this remarkable series of reactions were first worked out with yeast and muscle. The reasons were largely pragmatic. Yeasts had long been of interest as the producers of potable and industrial alcohol, and could readily be obtained and handled in large quantities. Muscle afforded the best opportunity of measuring the amount of work done in a biological system and correlating it with the respiratory turnover. Yeast consists of unicells which are capable of prolonged existence and multiplication without oxygen, and muscle of very specialised tissue cells with a high oxygen requirement. The fact that, contrary to all expectation, the series turned out to be identical in two such diverse types of cell suggested that it might well be universal. This has now been confirmed in considerable experimental detail for the most widely ranging cell types from bacteria and protozoa, to those of the various tissues of higher plants and animals. The evidence rests on the isolation of all the intermediates and requisite enzymes, and the inhibition of respiration, either anaerobic or aerobic, by iodoacetate and fluoride which specifically inhibit triosephosphate dehydrogenase and enolase respectively, two of the enzymes concerned. In view of its universality this breakdown of sugars to pyruvic acid may well be called normal glycolysis.

Normal glycolysis

The first important step towards an analysis was made when Harden and Young showed that, while an extracted yeast juice is breaking down glucose, additions of inorganic phosphate cause temporary increases in the rate of reaction. They showed that the phosphate is locked up in fructofuranose-1,6-diphosphate (Harden and Young ester). This was in 1905 and over the next thirty years the work of numerous outstanding biochemists completed the story. The series of reactions is sometimes called the Embden-Meyerhof pathway in honour of two of them.

Most cells contain a carbohydrate pool, i.e. a mixture of sugars and sugar anhydrides (e.g. sucrose and starch in plants, and glycogen in yeasts and animals) which, under the influence of the cells' enzymes, are readily inter-convertible. Here we need only concern ourselves with the manner in which respiration draws glucose units from the pool.

The whole process, glucose → pyruvic acid, may be considered for convenience, but, of course, quite artificially, in four stages. (1) Phosphorylation of glucose to fructofuranose-1,6-diphosphate; (2) splitting of the 6C hexose chain to two 3C triose units; (3) anaerobic oxidation of glyceraldehyde-3-phosphate to 1,3-diphosphoglyceric acid; (4) dephosphorylation of 1,3-diphosphoglyceric acid to pyruvic acid.

The formation of fructofuranose-1,6-diphosphate
Starting with glucose itself the series of reactions can be represented:

4)   glucose + ATP $\underset{}{\overset{hexokinase}{\rightleftharpoons}}$ glucose-6-phosphate + ADP

5) glucose-6-phosphate $\underset{\overset{\xrightarrow{\hspace{2cm}}}{\xleftarrow{\hspace{2cm}}}}{\overset{\textit{phosphohexoisomerase}}{}}$ fructose-6-phosphate

6) fructose-6-phosphate + ATP $\underset{\overset{\xrightarrow{\hspace{2cm}}}{\xleftarrow{\hspace{2cm}}}}{\overset{\textit{phosphofructokinase}}{}}$ fructose-1,6-diphosphate + ADP

In these reactions two molecules of adenosine triphosphate (ATP) are dephosphorylated to adenosine diphosphate (ADP). Starch and glycogen, however, can be converted to glucose-1-phosphate by the appropriate phosphorylases using inorganic phosphate only. The glucose-1-phosphate may be transformed by phosphoglucomutase to glucose-6-phosphate, so that the formation of fructose diphosphate from starch or glycogen involves only a single ATP molecule.

Splitting the carbon chain

In the presence of the enzymes, aldolase and triosephosphate isomerase, fructofuranose-1,6-diphosphate is converted to an equilibrium mixture with dihydroxyacetone phosphate and glyceraldehyde-3-phosphate. Writing phosphoric acid as $HO\,\circledP$ the reaction is

7)
$$\text{fructose-1,6-diphosphate}$$

$$CH_2OH\ CO\ CH_2O\,\circledP \xleftrightarrow{\textit{triosephosphate isomerase}} CHO\ CHOH\ CH_2O\,\circledP$$

dihydroxyacetone phosphate   glyceraldehyde-3-phosphate

and equilibrium lies far towards the left; but it is the glyceraldehyde-3-phosphate that is continuously oxidised.

Anaerobic oxidation of glyceraldehyde-3-phosphate

The oxidant is nicotinamide adenine dinucleotide (NAD p. 37). It was discovered by Warburg that a crystalline preparation of triosephosphate (glyceraldehyde-3-phosphate) dehydrogenase would not bring about oxidation unless inorganic phosphate was added and that the product was a diphosphate.

8)
$$
\begin{array}{c}
CHO \\
| \\
CHOH \\
| \\
CH_2O\,\circledP
\end{array}
+ NAD + H_3PO_4 \xleftrightarrow{\hspace{1cm}}
\begin{array}{c}
COO\,\circledP \\
| \\
CHOH \\
| \\
CH_2O\,\circledP
\end{array}
+ NADH_2
$$

glyceraldehyde-3-phosphate   1,3-diphosphoglyceric acid

The oxidation and phosphorylation occur together at the surface of the enzyme and neither can occur without the other. This reaction is the classic example of a strictly linked oxidation-phosphorylation and still the best documented; but what exactly goes on at the enzyme surface is still open to discussion. The enzyme is particularly sensitive to inhibition by thiol reagents, such as iodoacetate, perhaps because it is bound to the NAD by way of —SH groups of glutathione. The reagents remove the NAD.

Dephosphorylation of 1,3-diphosphoglyceric acid
The loosely bound phosphate at the number 1 position is transferred to ADP by the enzyme phosphoglycerylkinase to give the relatively stable 3-phosphoglyceric acid and ATP.

9) $COO®$ $CHOH$ $CH_2O®$ $+ ADP \xrightarrow{\text{phosphoglyceryl kinase}}$ $COOH$ $CHOH$ $CH_2O®$
$+ ATP$

Phosphoglyceromutase then catalyses a conversion of the acid to 2-phosphoglyceric acid, possibly through a 2,3-diphosphoglyceric intermediate.

10) $COOH$ $CHOH$ $CH_2O®$ $\xrightarrow{\text{phosphoglyceromutase}}$ $COOH$ $CHO®CH_2OH$

This is followed by a dehydration to phosphoenolpyruvate (PEP). The enolase responsible requires magnesium and is strongly inhibited by fluoride.

11) $COOH$ $CHO®CH_2OH \xrightarrow{\text{enolase}} COOH$ $CO®=CH_2 + H_2O$

In the last step, pyruvic kinase transfers the remaining phosphate group to ADP. The reaction is reversible; but under normal circumstances the equilibrium lies very far towards the (keto) pyruvic acid side.

12) $COOH$ $CO®=CH_2 + ADP \xrightarrow{\text{pyruvic kinase}} COOH$ $C=O$ $CH_3 + ATP$

Summing the reactions 4) to 12) and remembering to double 8) and onwards gives

13) $C_6H_{12}O_6 + 2\ NAD + 2\ ADP + 2® = 2\ C_3H_6O_3 + 2\ NADH_2 + 2\ ATP$

i.e. at the expense of reducing 2 NAD (transferring 4H or 4e) there has been a net gain of 2 ATP. If the source of glucose units is a polysaccharide there may be a gain of 3 ATP, without additional reduction of NAD. These two substances play key roles in the series of reactions: NAD as the only oxidant ('releaser' of energy) and ATP as the only conserver of energy, preventing its immediate dissipation during the oxidation. They play these roles, not only in the glycolysis just described, but in a vast range of metabolic sequences and cycles, including those of aerobic respiration and photosynthesis.

NAD and oxido-reduction

Nicotinamide adenine dinucleotide has the structure

adenine
|
Ⓟ — ribose
|
Ⓟ — ribose
|
nicotinamide

set out more fully in Figure 15, p. 57.

Its reduction and reoxidation resembles that of nicotinamide methiodide and is limited to the nicotinamide part of the molecule. Vennesland's experiments with ethanol and highly purified yeast alcohol dehydrogenase have shown, moreover, that H atoms (not electrons) are transported.

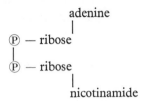

14)

| ethanol | NAD | acetaldehyde | NADH$_2$ |

In neutral solution the nitrogen becomes trivalent and NADH$_2$ is strictly NADH + H$^+$. Deuterium-labelled ethanol, CH_3CD_2OH, gives NADH$_2$ labelled at the 4 position with 1 atom only as indicated by the asterisks above. Reoxidation of the labelled nucleotide by acetaldehyde gives CH_3CDHOH, i.e. the same D (or H in an unlabelled system) is transferred and the system is stereospecific: one of the H atoms at position 4 is placed above the nicotinamide ring and one below. Similar methods cannot be applied to the H exchange between the hydroxyl and nitrogen because exchange with the hydrogen ions of the medium occurs non-enzymically at both these positions.

The triosephosphate dehydrogenase active in glycolysis catalyses H transfer to NAD in a similar fashion; but there is one important difference, its reaction is with the other side of the nicotinamide ring. This means that NAD may be attached to two dehydrogenases simultaneously and that H can be shuttled from one to the other without ever leaving the enzyme surface, as indicated in Figure 8.

Enzymes with the same specificity as alcohol dehydrogenase are said to have A specificity and include lactic, malic and glycerate dehydrogenases; those with B specificity include triosephosphate (glyceraldehyde-3-phosphate),

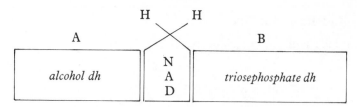

Figure 8 Diagram to show A and B specificity of NAD.

α-glycerophosphate and glutamic dehydrogenases, and NADH$_2$-cytochrome c reductase.

NADP, nicotinamide adenine dinucleotide phosphate, differs from NAD only in having a third phosphate group which is attached to the ribose ring remote from the nicotinamide. NADP has redox properties like those of NAD; but dehydrogenases act specifically with one or the other, not with both. For example, glucose-6-phosphate dehydrogenase transfers H to NADP in the respiratory reactions described on p. 68. Moreover, there are two distinct glyceraldehyde-3-phosphate dehydrogenases requiring phosphates; the one already mentioned as moving H to NAD in glycolysis and another catalysing H transfers with NADP in photosynthesis. The last two dehydrogenases are probably structurally separated within the cell (p. 130); but, where such separations do not exist, this sort of specificity may prevent reaction sequences interfering with one another; e.g. the oxidation of glucose-6-phosphate will not draw off NAD required for the oxidation of glyceraldehyde-3-phosphate.

It is possible that no cells are without NAD and NADP; other cell oxidants to be mentioned later are certainly more restricted in their distribution. This may lead to the idea that NAD and the glycolytic sequence were evolved at a very early stage; they could have antedated the advent of oxygen in the atmosphere since the latter may be of biological origin. They are, however, so chemically sophisticated, that some simpler methods of energy transfer must presumably have come before them.

ATP and high-energy compounds

Adenosine triphosphate holds a similar position on the side of energy conservation. It appears to be universally present in cells, and similar nucleotides, like uridine triphosphate, have much more limited occurrence and functions.

Coupling the formation of anhydride bonds to an oxidation 'fixes' or 'conserves' energy, i.e. retains some part of the energy of reaction within the system instead of allowing it to be scattered as in a combustion. For biological purposes the energy level represented by the products of complete combustion, $CO_2 + H_2O$, is taken as a reference level, and a high energy compound is one with a relatively large heat of combustion—such as fat or carbohydrate.

In 1941 Lipmann deduced that, in certain phosphate reactions, the free energy difference between reactants and products was not as great as would have been expected. He therefore proposed that the 'missing' energy was retained in the phosphate bonds over and above the 3–4 kcal usual in the formation of normal phosphate-ester linkages. He proposed 4 types of such special phosphate linkage and others have since been added. Three of them are involved in glycolysis:

Enolphosphate, e.g. phosphoenolpyruvate

$$CH_2=CO \sim \textcircled{P}COOH$$

Carboxylphosphate, e.g. 1,3-diphosphoglyceric acid

$$COO \sim \textcircled{P}CHOH\ CH_2O\,\textcircled{P}$$

Pyrophosphate, e.g. adenosine triphosphate

$$Adenosine\ \textcircled{P} \sim \textcircled{P} \sim \textcircled{P}$$

where $\sim \textcircled{P}$ represents the specially linked phosphate. Lipmann termed these particular bonds 'energy rich', since a relatively large amount of energy is released when they are hydrolysed. This is in conflict with physico-chemical terminology in which a high energy bond is one requiring a large amount of energy to break it; but the Lipmann usage is convenient and common in biochemistry. Lipmann estimated the average loss of free energy when a $\sim \textcircled{P}$ linkage is hydrolysed as about -12 kcal per mole under standard conditions. More recent calculations with improved data reduce this standard value (ΔG_0) to around -9 kcal. At the much lower concentrations actually present in cells, temperatures around $25°C$ and pH 6–7 (instead of pH 0) ΔG may rise to -12 or -16 kcal.

Adenosine triphosphate may be written, adenosine $\textcircled{P} \sim \textcircled{P} \sim \textcircled{P}$. The adenosine diphosphate (ADP) which remains after the terminal $\sim \textcircled{P}$ has been removed still has an energy rich link and its dephosphorylation to adenosine monophosphate (AMP) with ΔG_0 also about -9 kcal may occur, though not apparently in glycolysis. Pyrophosphate splitting of ATP may also lead to the formation of AMP direct. Since ADP is the most usual phosphate acceptor it is important that cells often contain an enzyme, adenylate kinase which catalyses

15) $AMP + ATP \rightleftharpoons 2\,ADP$

which, so to say, brings the adenosine nucleotide back into normal circulation.

Conversely, working from right to left the reaction, can provide muscle cells with a temporary supply of ATP, when conditions favour reaction in that direction.

The use of the term 'high energy bond' and the symbol $\sim \textcircled{P}$ to represent it may be misleading. The extra energy results from the sum of the intramolecular changes that occur when such a bond is broken and the freed valencies are taken up by the elements of water. In other words, the energy-rich state is not a characteristic of the bond in itself, but is conferred upon it by the molecular structure of which it forms a part. If, however, we speak instead of energy-rich compounds we risk confusion with substances such as the carbohydrates which have large stores of free energy that are available even if not so readily.

The limitations of glycolysis

Glycolysis taken in isolation and as summarised by the equation

16) $C_6H_{12}O_6 + 2\,NAD + 2\,ADP + 2\textcircled{P} = 2\,CH_3COCOOH + 2\,NADH_2 + 2\,ATP$

is unsatisfactory as a continuing biological process. It has three serious shortcomings. The continued accumulation of pyruvic acid would soon become toxic; NAD is always in short supply (e.g. about 6×10^{-5} M in *Sauromatum* spadix and about 8×10^{-4} M in yeast, exceptionally active subjects) so that its complete reduction to $NADH_2$ would soon bring the process to a stop; and finally the free energy gradient is inadequate. The standard free energy changes of all the partial reactions have been calculated. Taking into account the net formation of ATP, summing them algebraically gives an overall value $\Delta G_0 = +\,1\cdot2$ kcal. Introducing corrections for the presumed low reactant concentrations in living cells would perhaps convert this to a low negative value.

There are clearly numerous ways in which the situation can be biologically improved. Muscle cells, for example, start with glycogen, which gives glucose 6-phosphate with $\Delta G_0 = -1\cdot2$ kcal instead of the $+\,3\cdot9$ kcal required for its formation from glucose, an improvement of $5\cdot1$ kcal. But this does not solve the problem of the shortage of oxidant. *T. gambiense* and its allies, even though they do not oxidise pyruvic acid, do consume oxygen, and it is evident that they use it to reoxidise $NADH_2$ to NAD and so keep the glycolytic sequence going. When not actively multiplying they presumably excrete the pyruvic acid into the blood stream of the host. But, of course, by far the most effective improvement upon glycolysis as a respiratory mechanism, both chemically and thermodynamically, is the full aerobic oxidation of the pyruvic acid, which still contains more than four fifths of the available energy of the carbohydrate source. But short of this, the $NADH_2$ can be reoxidised anaerobically by partially oxidised materials within the cell. *T. gambiense* survives in oxygen-free blood; but then glycerol appears and the amount of pyruvic acid is reduced. If the pyruvic acid itself is used as oxidant it also solves the problem of getting rid of its toxicity. The ways in which glycolysis is anaerobically converted into an open ended process with continued reoxidation of the $NADH_2$ are considered in the next chapter.

Further Reading

Baldwin, E. (1963) *Dynamic Aspects of Biochemistry*, 4th ed. Cambridge University Press, Cambridge.
Burton, K. and Krebs, H. A. (1953) The free-energy changes associated with the individual steps of the tricarboxylic acid cycle, glycolysis and alcoholic fermentation and with the hydrolysis of the pyrophosphate groups of adenosine-triphosphate. *Biochem. J.*, **54**, 94–107.
Dagley, S. and Nicholson, D. E. (1970) *An Introduction to Metabolic Pathways*. Blackwell Scientific Publications, Oxford.
Oparin, A. I. (1961) *Life: its Nature, Origin and Development*. English translation by Ann Synge. Oliver and Boyd, Edinburgh.
Ramsay, J. A. (1965) *The Experimental Basis of Modern Biology*. Cambridge University Press, Cambridge.
Ryley, J. F. (1962) Studies on the metabolism of the Protozoa. 7 and 9 *Biochem. J.*, **62**, 215–24 and **85**, 211–23.

3 Respiration Without Oxygen

Lactic acid formation

It has been known for a long time that, when muscle cells are short of oxygen, as in violent exertion, they accumulate lactic acid, which is slowly removed in the blood stream or reoxidised during rest. Temporarily they are anaerobic to the extent that their energy demands are outstripping the possibilities of oxygen supply and glycolysis exceeds oxidation. The conversion of pyruvic acid to lactic acid represents perhaps the simplest way of making glycolysis open ended. It occurs in muscle cells because they possess the enzyme lactic dehydrogenase which catalyses the reaction.

17) $$CH_3CO\ COOH + NADH_2 \underset{\text{lactic dh}}{\overset{}{\longleftrightarrow}} CH_3CHOH\ COOH + NAD$$

The dehydrogenases as a class are 'soluble', i.e. readily brought into an aqueous sol outside the living cell without losing their activity. Since their action as catalysts must be reversible, it should in theory be possible to link the activities of any pair of dehydrogenases that use NAD as hydrogen acceptor. For example:

18)

pyruvic acid
```
    COOH
    |
    CO
    |
    CH₃
```

glyceroyl-1,3-diphosphate
```
    COO Ⓟ
    |
    CHOH
    |
    CH₂O Ⓟ
```

NADH₂

NAD

lactic acid
```
    COOH
    |
    CHOH
    |
    CH₃
```

glyceraldehyde-1,3-diphosphate
```
    CHO HO Ⓟ
    |
    CHOH
    |
    CH₂O Ⓟ
```

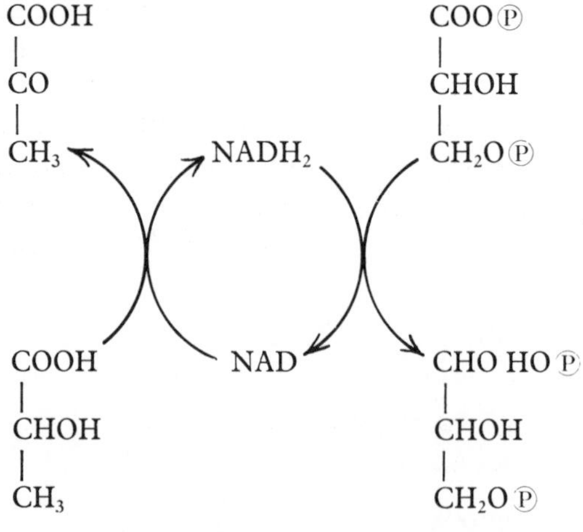

lactic dh   *triosephosphate dh*

leading in solution to an equilibrium mixture of the four reactants. This particular set of reactions may also be linked outside the cell via the glyceroyl-1,3-diphosphate to ATP formation from ADP. It occurs particularly readily because lactic dehydrogenase has A specificity with respect to NAD (p. 37) whereas triosephosphate dehydrogenase has B specificity. Inside the cell, so long as a supply of glucose units is available to be glycolysed to phosphoglyceraldehyde and on to pyruvic acid the process may be continuous, as outlined in Figure 9.

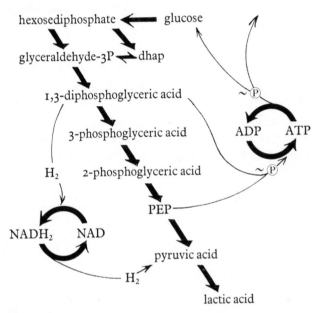

Figure 9   Anaerobic formation of lactic acid from glucose.

The cells of practically all animal tissues are said to be capable of lactic acid formation; kidney cells and erythrocytes form appreciable amounts even aerobically. Many plant cells form it anaerobically; though not as the sole product. Among microorganisms, it is a normal anaerobic product in *Lactobacillus* spp., *Streptococcus lactis*, protozoa such as *Tetrahymena pyriformis* and algae such as *Chlorella*. In the last, however, the product is D-lactic acid instead of the usual L isomer.

Alcohol formation

The production of alcohol from the reserve carbohydrates of plant storage tissues (potatoes, grapes, barley grains, etc.) is one of the oldest technologies. It may occur spontaneously; but although plant cells are capable of carrying out the complete series of reactions, it is much accelerated by the presence of yeasts. In most cases the plant tissue converts the insoluble carbohydrates to sugars, and the yeast takes over the glycolysis. In grapes, the carbohydrates are stored as soluble sugars and the epiphytic wine yeasts (varieties of *Saccharomyces ellipsoideus*) can get to work as soon as the surface is broken by over-ripening or crushing. There are many collateral processes such as give rise to the difference between a vintage wine and a poteen whisky; but the alcohol production is strictly from pyruvic acid formed by normal glycolysis.

The suppression of lactate in favour of alcohol formation is due to the presence of pyruvic decarboxylase, which usually secures the pyruvic acid even when lactic dehydrogenase is present. In extracts the reaction

$$19) \qquad CH_3CO\ COOH \xrightarrow[\textit{decarboxylase}]{\textit{pyruvic}} CH_3CHO + CO_2$$

goes virtually to completion. The enzyme from yeast is very active and that extracted from some higher plants under comparable conditions even more so. On the other hand, some plant tissues (many leaves and potatoes) may show little activity or none. In common with other keto acid decarboxylases the enzyme has diphosphothiamine (thiamine pyrophosphate) linked to its protein

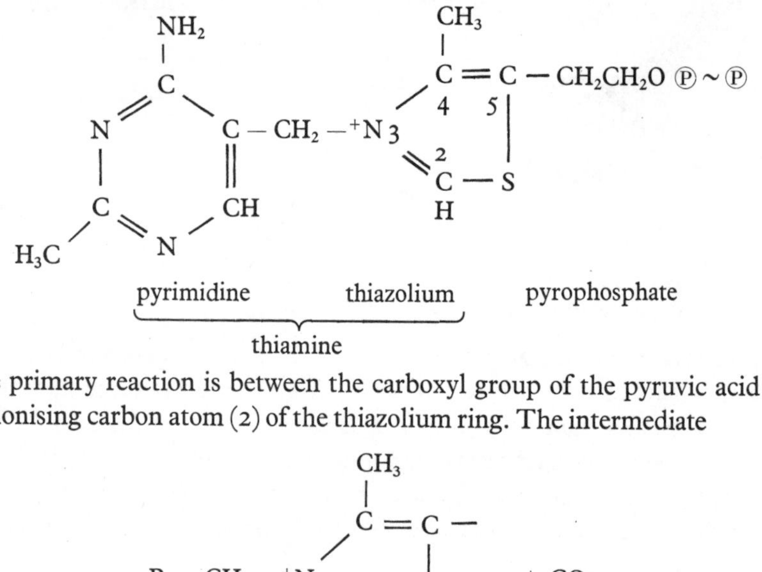

pyrimidine   thiazolium   pyrophosphate

thiamine

The primary reaction is between the carboxyl group of the pyruvic acid and the ionising carbon atom (2) of the thiazolium ring. The intermediate

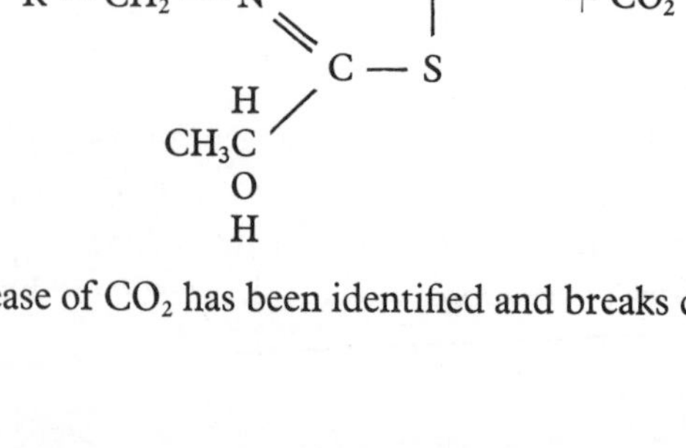

formed with release of CO_2 has been identified and breaks down to release the

CH₃CHO and restore the thiazolium. Mg^{2+} is a required cofactor and probably serves to link the thiamine pyrophosphate to the enzyme protein.

Acetaldehyde is reduced to ethanol, by an oxidation of glyceraldehyde diphosphate coupled through NAD in the same way that pyruvic acid is reduced to lactic (Eqn 18). Alcohol dehydrogenase, which has been obtained from yeast in a high state of purity, replaces the lactic dehydrogenase.

Carrot tissues are possibly the most single minded in anaerobic alcohol production. It has been shown that their anaerobic respiration satisfies the Gay–Lussac equation

20)
$$C_6H_{12}O_6 \longrightarrow 2C_2H_5OH + 2CO_2$$

to within 1 per cent and that no appreciable quantities of plant acids, glycerol, acetaldehyde or phosphate esters accumulate. This is probably unusual and even in yeasts there are measurable side products. At least one Metazoon, the acanthocephalan *Moniliformis dubius*, is also known to form alcohol and CO_2 in equimolecular amounts and in high yield. In anaerobic incubations, for each 6 molecules of alcohol and 6 molecules of CO_2, about 1 molecule of lactic acid and traces of succinic, acetic and butyric acids were recovered. The fact that an infection of *M. dubius* displaces tapeworms from the anterior quarter of rat intestines has been attributed to the excreted alcohol and considered to confer a biological advantage on the *Moniliformis*.

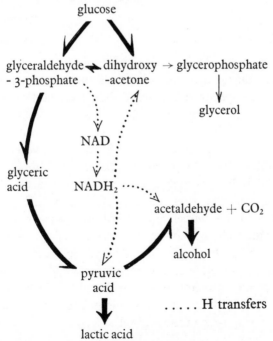

Figure 10 The principal anaerobic products of glucose.

Figure 10 illustrates the relations between the two anaerobic processes. It also indicates a third possibility.

Glycerol formation

It has already been mentioned that *Trypanosoma gambiense*, when deprived of air, forms glycerol from glucose with a simultaneous reduction in the amount of pyruvic acid. Other trypanosomes form a mixture of glycerol and pyruvic acid, even when aerated. The same also happens, though to a smaller extent, with yeast; Pasteur showed that glycerol normally accounts for 2–3 per cent of the sugar used. This results from an oxidoreduction between the two triosephosphates (Eqn 7, p. 35). Dihydroxyacetone phosphate, predominance of which is favoured by the equilibrium, acts as hydrogen acceptor instead of pyruvic acid (lactate formation) or acetaldehyde (alcohol formation). The hydrogen transfer occurs through NAD just as when pyruvic acid or acetaldehyde is the oxidant and the reduction is represented by Eqn 21; but the reaction in extracted systems is much slower and in most cells is more or less completely suppressed.

$$21) \quad \begin{array}{c} CH_2OH \\ | \\ CO \\ | \\ CH_2O\,\circledP \end{array} + NADH_2 \underset{dehydrogenase}{\overset{L\text{-}\alpha\text{-}glycerophosphate}{\rightleftarrows}} \begin{array}{c} CH_2OH \\ | \\ CHOH \\ | \\ CH_2O\,\circledP \end{array} + NAD$$

dihydroxyacetone phosphate   α-glycerophosphate

A soluble α-glycerophosphate dehydrogenase has been extracted from protozoan, insect, mammalian and other cells and may be responsible for the reaction. It has the B type of specificity for NAD, the same as triosephosphate dehydrogenase, and so cannot form a H-transporting unit with it as the lactic and alcohol dehydrogenases can. This may contribute to its relative ineffectiveness in the cell. Curiously enough its presence in yeast has been more difficult to establish. The α-glycerophosphate is hydrolysed to glycerol + phosphate, probably by a phosphatase. The reverse process of glycerol consumption probably has an entirely different enzyme (p. 128).

If the pyruvic acid or acetaldehyde concentration is reduced, e.g. by poisoning glycolysis with sodium fluoride, the amount of glycerol formed tends to increase. Neuberg 'fixed' the acetaldehyde in yeast as the bisulphate compound and the result was the formation of glycerol in good yield: approximately 1 mole glycerol to 1 mole acetaldehyde fixed. Alternatively, maintaining the yeast cultures at an alkaline pH gets rid of the acetaldehyde by a dismutation to acetic acid and alcohol.

$$22) \quad 2CH_3CHO + H_2O \longrightarrow CH_3COOH + C_2H_5OH$$

These processes were used in Germany during the acute fat shortage of the First World War to produce glycerol for explosives. They have also been used in this country; but it is normally cheaper to produce glycerol from vegetable fats, such as palm oils, as a side product of soap making.

Other end products

With so versatile a substance as pyruvic acid as a starting point, it is scarcely surprising that anaerobiosis may lead to numerous other end products, all representing an overall reduction. Some occur in small yield; but others may account for a large part of the carbohydrate consumed. The latter are confined to special groups or species of bacteria. As an example may be quoted the formic acid derived from decarboxylation of pyruvic acid by coenzyme A (p. 56).

23) $CH_3COCOOH + CoASH \rightleftharpoons CH_3COSCoA + HCOOH$
         pyruvic acid   CoA   acetyl CoA   formic acid

The formic acid may accumulate as end product (*Salmonella typhi*); but in the common intestinal *Escherichia coli* it is split by a hydrogenylase system to $H_2 + CO_2$. *E. coli* also converts the acetyl CoA through acetyl phosphate to acetic acid. Acetic acid is produced by some green algae probably by a similar mechanism. Numerous other degradations of pyruvic acid are detailed in standard text books of bacteriology. Some, such as butyl and isopropyl alcohol formation, have been used industrially. The ripening of Emmenthaler cheese is due to a propionic acid fermentation and the holes are caused by the simultaneous production of CO_2.

Non-glycolytic pathways

Leuconostoc mesenteroides, a bacterial coccus, converts glucose anaerobically to an equimolecular mixture of lactic acid, alcohol and CO_2. It was at first supposed that this was due to a normal glycolysis, with half the pyruvate being converted to CO_2 + alcohol and half being reduced to lactate. It was later found, however, that, when the bacterium was fed glucose—$1^{14}C$ (i.e. glucose labelled at the number 1 position), all the label was recovered in the CO_2 instead of in the alcohol or lactate as after normal glycolysis. Moreover, the bacterium was found not to possess aldolase and triosephosphate isomerase, two of the normal glycolytic enzymes. It contains instead glucose-6-phosphate dehydrogenase, an enzyme first discovered in yeast and red blood cells by Warburg and which is now known to occur very widely. It normally transfers 2H from glucose-6-phosphate to NADP; but the *Leuconostoc* version of it can also use NAD as H-acceptor. The oxidation product,

6-phosphogluconate, is further oxidised by 6-phosphogluconic dehydrogenase with loss of CO_2 from the number 1 position to give ribulose-5-phosphate.

24) $CH_2O\textcircled{P}(CHOH)_4CHO + NAD + H_2O \rightleftharpoons CH_2O\textcircled{P}(CHOH)_4\ COOH +$
$$NADH_2$$

glucose-6-phosphate   6-phosphogluconate

25) $CH_2O\textcircled{P}(CHOH)_4\ COOH + NAD \rightleftharpoons CH_2O\textcircled{P}(CHOH)_2\ COCH_2OH +$
$$CO_2 + NADH_2$$

6-phosphogluconate   ribulose-5-phosphate

The ribulose-5-phosphate is converted to xylulose-5-phosphate and split with uptake of P_i into glyceraldehyde-3-phosphate and acetyl phosphate. The glyceraldehyde-3-phosphate is converted to lactate by the usual glycolytic reactions. The acetyl phosphate is reduced to alcohol by acetaldehyde and alcohol dehydrogenases. These two enzymes both use NAD so that the two initial oxidations are linked to the two reductions, as indicated in Figure 11,

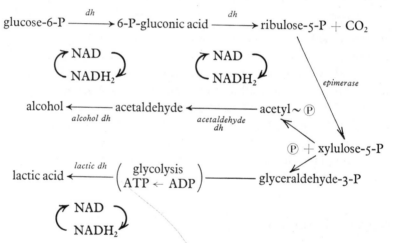

Figure 11 The hexosemonophosphate shunt in *Leuconostoc*. Enzymes are shown in italics.

and the process proceeds strictly anaerobically. The net gain of ATP molecules is only one instead of the two in normal glycolysis since the route to alcohol via acetyl does not yield any. As the reactions branch from the normal glycolytic pathway at glucose-6-phosphate, they are frequently called a hexosemonophosphate shunt. It will be seen later that similar shunts occur more frequently under aerobic conditions and through the agency of a glucose-6-phosphate dehydrogenase that is specific to NADP and so cannot link on directly to internal reductions depending on $NADH_2$.

Another type of hexosemonophosphate shunt (the Entner–Doudoroff) that is fully anaerobic is utilised by *Pseudomonas lindneri*. Glucose-6-phosphate is oxidised as before, (Eqn 24), but the 6-phosphogluconate formed instead of being decarboxylated is dehydrated to give 2-keto-3-deoxy-6-phosphogluconate

26) $CH_2O(P)(CHOH)_4COOH \rightleftharpoons CH_2O(P)(CHOH)_2 CH_2COCOOH + H_2O$

which is split by an aldolase to glyceraldehyde-3-phosphate and pyruvate.

27) $CH_2O(P)(CHOH)_2CH_2COCOOH \rightleftharpoons CH_2O(P)CHOH\ CHO + CH_3CO\ COOH$

The glyceraldehyde-3-phosphate is converted to pyruvate by the normal glycolytic reactions and the total pyruvate to CO_2 + alcohol (Figure 12).

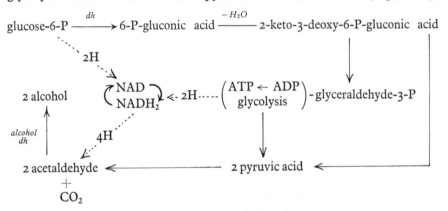

Figure 12   The hexosemonophosphate shunt in *Pseudomonas*.

There is again only half the ATP formation of normal glycolysis since one pyruvate molecule is produced without oxidation of glyceraldehyde-3-phosphate.

Energy turnover

Anaerobic respiration can proceed in a suitable cell system at measurable rates because there is an overall loss of free energy. This is principally because the later reactions of the series are highly exergonic, i.e. proceed with considerable loss of free energy; and because the successive reactions are united all along the chain by common reactants. On a hydrostatic analogy one is reminded of a siphon in which the water may at first rise because of the cohesion of its molecules, providing that an exit lower than the supply level ensures an overall loss of free energy. It will be noted, however, that the energy 'siphon' of anaerobic respiration is boosted by the injection of energy from ATP in the early stages.

Figure 13 illustrates the energy levels involved in the anaerobic respiration of glycogen → lactate and glucose → alcohol. In both series, four ~ (P) bonds are generated in ATP molecules. The lactate formation from glycogen in muscle 'borrows' one, leaving a net gain of 3. This represents the conservation of about $3 \times 9 = 27$ kcal (at standard conditions) out of 64 kcal released, or about 42 per cent. The alcoholic fermentation of yeast, starting with glucose, uses an additional $\sim$ ℗ in the formation of glucose-6-phosphate. The net conservation is therefore only $2 \times 9 = 18$ kcal out of the 56 kcal available, or 32 per cent. In the non-glycolytic pathways described it is even less. It cannot even be supposed that all the energy temporarily conserved in ATP will eventually be of service to the cell, since the further transformations in the formation of protein, structural carbohydrates etc. will also run at less than 100 per cent efficiency.

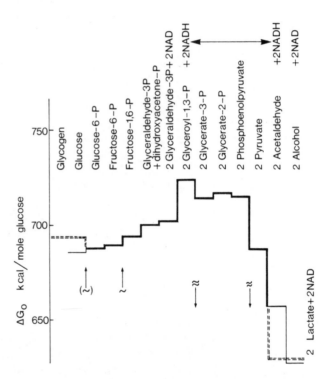

Figure 13 Energy levels in anaerobic respiration and glycolysis; = = = = glycogen ———→ lactic acid; ——— glucose to alchohol; ——— common to both processes. $\overset{\uparrow}{\sim}$ = $\sim$ ℗ taken up $\underset{\downarrow}{\approx}$ = $2 \sim$ ℗ released.

Short-term lack of oxygen

Most cells continue to respire when deprived of oxygen. Sperms, for example, find themselves under anaerobic conditions when deposited in the female reproductive tract. They maintain a high degree of mobility for up to about 20 minutes by the anaerobic respiration of fructose included in the seminal fluid.

Even such an aerobic system as nerve tissue continues to emit heat at a reduced rate and accumulate lactic acid for several hours. If there was a simple transition to lactate formation, there would be no emission of CO_2; but this does, in fact, continue for a while and, when nerve cells are brought back to air, the burst of oxygen uptake does not fully account for that missed during anaerobiosis.

Among lower animals the capacity to endure anaerobic conditions is more pronounced. For example, the cockroach, *Cryptocerus punctatus*, can respire anaerobically for periods of 5–6 hours. Lactic acid is formed and on return to air is fully oxidised. Intestinal parasites, such as tapeworms, live in a habitat which normally has an oxygen content around 5 per cent and their respiration is at least partially anaerobic. They contain large amounts of glycogen which, in the sheep tapeworm, *Moniezia expansa*, is converted to CO_2, lactic, succinic and fatty acids.

The nematode *Ascaris* has been studied in some detail. *Ascaris mystax* is able to live anaerobically for 5–6 days. Species like *A. lumbricus* live indefinitely in the host intestines and to a very large extent depend on their anaerobic respiration for their energy supply. They possess an oxidation system (described more fully on p. 88) which is virtually inactive at the low oxygen pressures of the intestinal gases. The main product of the anaerobic respiration is succinic acid accompanied by acetic, propionic and other acids in considerable variety. Lactic acid is not formed though the worm's glycogen is glycolysed by the normal pathway and linked with phosphorylation. The $NADH_2$ however is not reoxidised anaerobically by pyruvic acid itself; which, in the CO_2-rich intestinal atmosphere, is carboxylated giving rise to fumaric acid. This acts as the H_2-acceptor so giving the final succinic acid. Although the anaerobic respiration thus becomes open ended, it is doubted whether the parasite can live a normal life entirely without oxygen; but this is not necessarily due to any requirement of its muscle cells.

The cells of many plant storage tissues (carrots, potatoes, etc.) can endure much longer periods of anaerobiosis, days or weeks rather than minutes or hours. Tissues like carrots which form alcohol, continue to emit CO_2, often at rates exceeding that in air. As the CO_2-output in the transition glucose to alcohol is only one third of that in complete oxidation, it would appear that glucose is being consumed considerably faster under nitrogen. Figure 14 shows the course of CO_2 emission when carrot tissues are held for several days under nitrogen and then returned to air. Potato tissues show behaviour more like that of nerve cells, though on a more extended time scale. Under nitrogen their CO_2-output falls away gradually to almost nothing at all; but on return to air there is a burst of extra CO_2-emission before the rate returns to the air line. Potatoes normally show very little pyruvic decarboxylase activity and,

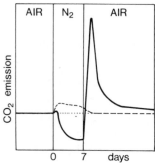

Figure 14 Course of CO_2-output by potatoes —————— and carrot slices – – – – in air and nitrogen. Initial rates adjusted to equality.

for about 7 days form no alcohol under nitrogen. They form instead a mixture of products of which lactate forms about half. The temporary burst of CO_2-emission on return to air is due to the oxidation of this and perhaps other products. There is no corresponding oxidation in carrots of the alcohol, whose formation appears to be irreversible. When carrot discs are immersed in a 5 per cent alcohol solution their CO_2-emission increases somewhat, but if the alcohol is labelled with ^{14}C, no label is recovered in the CO_2 emitted, or in any of the acids that are the intermediate stages of oxidation. Carrots kept under a current of nitrogen do not, however, accumulate high concentrations of alcohol, since it evaporates into the gas stream. All the above tissue cells can return to normal when brought back into air, providing that their nitrogen experience has not been excessively long.

Ecology of anaerobiosis

Cells embedded in bulky tissues may create their own anaerobic environment. This can happen even when they are provided with oxygen by a pressure-boosted system like the blood stream. The classic example is fatigued muscle as already mentioned. The lens tissue of eyes, which is not reached by the blood supply system, has a very low rate of metabolism and about 90 per cent of the glucose reaching it is glycolysed to lactate. Self-induced anaerobiosis can also occur in the centre of bulky plant tissues where oxygen supply depends on simple diffusion through the intercellular air spaces.

But the most notable auto-anaerobiosis is that of yeast. Under common brewing conditions, i.e. about 1 lb yeast to a barrel of wort (3 g/l) at room temperature, respiration is so rapid that all oxygen is removed in about 24 hours. The yeast, however, continues to ferment vigorously and, as Pasteur found, for a time goes on growing and dividing.

Anaerobic sites on the large scale are mostly limited to swamps and the lower layers of stagnant water generally. Their lack of oxygen is itself usually due to the respiration of micro-organisms of one kind or another. They may be colonised by annelids, such as *Tubifex* that can exist anaerobically for periods up to 60 hours, incurring an 'oxygen debt' in doing so. During the phase of anaerobic respiration their glycogen consumption may be four times as fast as in a normal atmosphere; but on returning to aerobic conditions much of the lactic acid they have formed is resynthesised to glycogen. Similar cycles can occur on a lesser scale in the common earthworm.

The most important anaerobic sites are rice fields, the surfaces of which are heavily colonised by blue-green algae. The rice plants themselves, like virtually all other swamp plants, evade, rather than endure, the anaerobic conditions round their roots. The respiration of their root cells is fully aerobic, depending on oxygen diffusing through the intercellular spaces from above. The aeration is so effective that, reversing the normal procedure, the surface cells of the roots oxidise the adjacent soil. Blue clays may act as indicators becoming red at the surface of the roots due to the oxidation of their iron salts. The underground stems of water lilies can colonise the anaerobic mud at the bottom of stagnant ponds. Air entering the floating leaves diffuses down the very porous leaf stalks, which can attain a length of 9 feet (3 metres), and the respiration of the root and stalk cells is fully aerobic. In winter, however, the leaves die and the air-line to the stems is cut off. They become dormant; but a slow respiration continues anaerobically and some alcohol accumulates in their cells.

Nitrate instead of oxygen

An altogether different type of evasion is practised by some bacteria, such as *E. coli* and the *Pseudomonas* group. When oxygen is absent but nitrate present they can reoxidise their $NADH_2$ by means of the nitrate. Similarly, *Desulphovibrio* cells reduce sulphate. These reactions depend on an electron transport system involving cytochromes and similar to these of normal aerobic respiration. Other oxidising agents may also replace atmospheric oxygen. Yeast, possessing a cytochrome *c* peroxidase, can respire more rapidly in the presence of external hydrogen peroxide.

Whatever may have been the state of affairs at the start of the biological era, anaerobic cells that can grow and multiply over an extensive period by using only metabolic intermediates to reoxidise their $NADH_2$ are nowadays a small minority. Fermentative bacteria like *Lactobacillus* and some *Clostridia* are notable examples. A few like *C. welchii* are actually poisoned by oxygen.

The inadequacy of anaerobic respiration

The great majority of cells fail to develop when restricted to anaerobic respiration, or even to survive for a normal period. It has long been known that cell division will not occur anaerobically, even though nuclear division will continue once it has begun. The germination of seeds, which involves numerous cell divisions, is also prevented by absence of oxygen. Rice grains were, for long, supposed to provide an exception; but rigorous examination has shown that they are capable only of a limited extension of certain preformed cells. If this brings the tip of the seedling into an aerated region germination can proceed; but under strictly anaerobic conditions there are no cell divisions and no synthesis of new protein occurs. The protoplasmic streaming of amoebae and other cells is also stopped. Active uptake of salts against normal concentration gradients depends on the presence of oxygen in such diverse cells as those of kidney and plant roots; and in its absence may even be reversed suggesting the breakdown of the normal membrane machinery. In short, in most cells activities depending directly or indirectly upon respiration fail in the absence of oxygen. Anaerobic respiration only exploits about a tenth of the free energy available in the aerobic respiration of a given amount of sugar, and it is often suggested that it therefore fails on purely quantitative grounds. As an explanation this is by itself inadequate. To begin with, some organisms do grow vigorously on a strictly anaerobic respiration based on normal glycolysis. *Lactobacillus casei* grows and multiplies vigorously on a glucose medium until the lactic acid concentration becomes toxic. In yeast and many other cells the rate of consumption of glucose may be several times faster in nitrogen than in air and the rate of energy turnover is greatly in excess of any ascertainable need of the cell. It must therefore be supposed that the ATP formation demonstrably occurring during anaerobiosis is not capable of being linked to at least some of the vital requirements of the cell; there is a 'failure of communication'. It is significant that poisons such as KCN that inhibit various stages of oxidative respiration without affecting the anaerobic phase also inhibit cellular salt uptake, protoplasmic streaming and so on. Unfortunately we know very little at present about how aerobic respiration is geared to ATP formation and utilisation.

Further Reading

Bishop, M. W. H. and Austin, C. R. (1957) Mammalian Spermatozoa. *Endeavour,* **16,** 137–50.

Bueding, E. (1963) Electron transport and fermentations in *Ascaris lumbricoides,* in *Control Mechanisms in Respiration and Fermentation.* Ronald Press, New York.

Gunsalus, I. C. and Stanier, R. Y. (1961) *The Bacteria,* 2nd ed. *Vol. II: Metabolism.* Academic Press, New York.

James, W. O. and Ritchie Ann F. (1955) The anaerobic respiration of carrot tissue. *Proc. Roy. Soc.,* **B 143,** 302–10.

Lehninger, A. L. (1965) *Bioenergetics.* Benjamin, New York.

Oginsky, E. L. and Umbreit, W. W. (1959) *An Introduction to Bacterial Physiology*, 2nd ed. Freeman, San Francisco and London.

Rose, A. H. (1968) *Chemical Microbiology*. 2nd ed. Chapter 6. Butterworths, London.

Stanier, R. Y., Doudoroff, M. and Adelberg, E. A. (1971) *General Microbiology*. 3rd ed. Macmillan, London.

Ward, P. F. V. and Crompton, D. W. T. (1969) The alcoholic fermentation of glucose by *Moniliformis dubius* (Acanthocephala), *in vitro*. *Proc. Roy. Soc.*, **B 172,** 65–88.

4 Oxidative Decarboxylation and Exergonic Carboxylation

Decarboxylation of pyruvic acid

Most cells, perhaps the great majority, continue to glycolyse in air and pyruvic acid goes on being formed. Given adequate aeration, it is not, however, reduced to lactic acid, or decarboxylated with formation of alcohol. Nor is it oxidised through a simple linear series of intermediates to carbon dioxide and water. Instead, it is oxidatively decarboxylated in a series of closely-knit reactions, which probably occur at the surface of a single enzyme complex with a particle weight around 4×10^6. No intermediate compounds are released and this has naturally not simplified the elucidation of what actually goes on.

In this situation, progress has been made largely by studying the co-factors required; these are at least five: thiamine pyrophosphate (TPP), Mg^{2+}, α-lipoic acid (lip S_2) coenzyme A (CoA or CoASH) and NAD.

The requirement for thiamine pyrophosphate and Mg^{2+} ions is similar to that of the anaerobic reaction (p. 44). The importance of α-lipoic acid emerged largely from work with bacterial cells, such as *E. coli*, *Lactobacillus casei*, etc. It is 6, 8-dithio-n-octanoic acid and reduction of the disulphide link gives dihydrolipoic acid,

28)

$$
\begin{array}{ccc}
& CH_2 & \\
H_2C & CH\,(CH_2)_4\,COOH & \\
| & | & \\
S\text{——}S & &
\end{array}
\qquad
\underset{\longrightarrow}{\overset{2H}{\rightleftharpoons}}
\qquad
\begin{array}{ccc}
& CH_2 & \\
H_2C & CH(CH_2)_4\,COOH & \\
| & | & \\
SH & SH &
\end{array}
$$

$\alpha-$ lipoic acid   dihydrolipoic acid

Coenzyme A is 3-phospho-adenosinediphosphate pantetheine (Figure 15). It is therefore similar to NAD and NADP with their ribosyl nicotinamide replaced by pantetheine. But the active group of NAD and NADP lies in

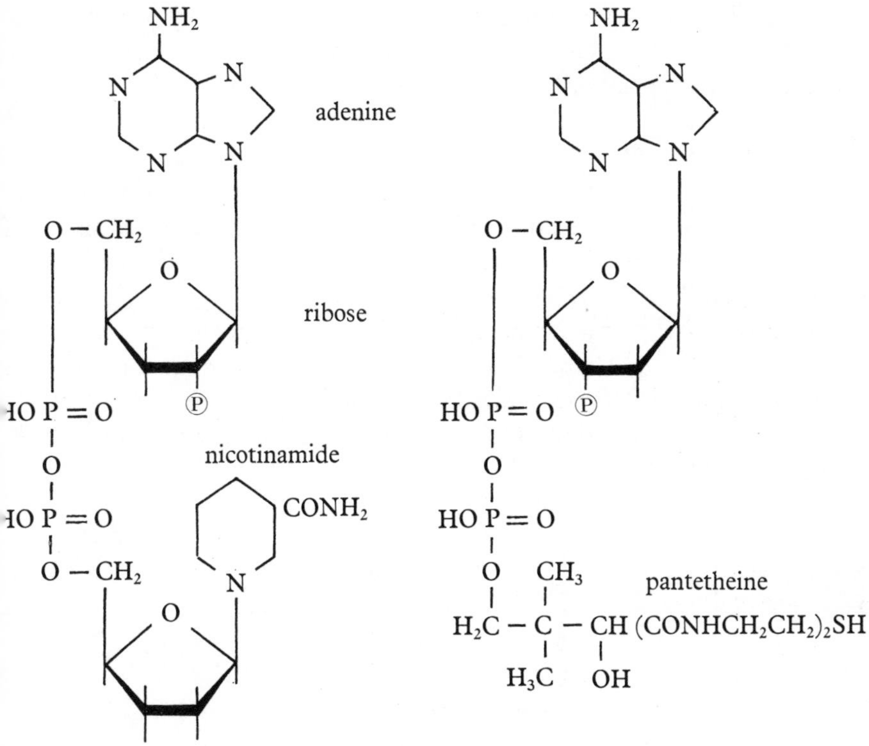

Figure 15 Formulae of reduced nicotinamide adenine dinucleotide phosphate (left) and coenzyme A (right).

nicotinamide and the active group of CoA is the terminal thiol group of the pantetheine. The NAD acts as a hydrogen transporter and the CoA as an acyl (RCH_2CO-) transporter.

The overall reaction with pyruvic acid can be summarised

29) $CH_3CO\ COOH + CoASH + NAD \rightarrow CH_3COSCoA + CO_2 + NADH_2$

Tentatively, the catalytic requirement for thiamine pyrophosphate and lipoic acid can be accounted for in outline by the following sequence. The Mg^{2+} is presumed to keep the thiamine pyrophosphate attached to the carboxylase fraction of the enzyme complex.

30)   Pyruvic acid + TPP $\xrightarrow[Mg^{2+}]{carboxylase}$ acetaldehyde TPP + CO_2

31)   Acetaldehyde TPP + lip S_2 → acetyl S lip SH + TPP

32)   Acetyl S lip SH + CoASH → acetyl SCoA + lip SH_2

33)   lip SH_2 + NAD → lip S_2 + $NADH_2$

Some such sequence is supposed with varying degrees of probability to occur in most animal tissues and many bacteria. There is some evidence of similar reactions in plant cells.

The formation of acetyl CoA

The elucidation of the structure and behaviour of coenzyme A has been as great a contribution to the study of metabolism as was that of coenzyme I (now NAD). It has been the achievement of numerous brilliant biochemists during the past thirty years and is still continuing. The central importance of CoA lies in the fact that it can accept acyl groups from many sources (just as the 2H accepted by NAD has numerous origins) and donate them to various acceptors (Figure 16). Moreover, the acyls transported are not limited to acetyl but include propionyl, CH_3CH_2CO-, succinyl, $HOOCCH_2CH_2CO$-, and many others.

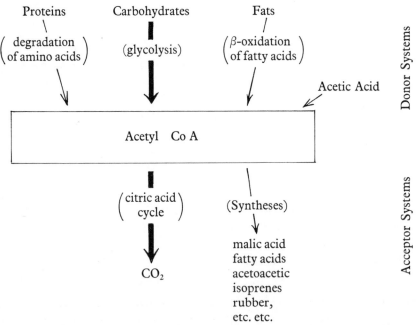

Figure 16   The principal metabolic relationships of acetyl CoA.

The citric acid cycle

So far as respiration is concerned the most important donor of CH_3CO- to CoA is pyruvic acid, and the most important acceptor is oxaloacetic acid with which it condenses to initiate the citric acid (Krebs, tricarboxylic or TCA) cycle. This cycle of reactions occupies a central position not only in cell respiration, but in cellular metabolism as a whole. Into it are channelled the intermediates of carbohydrate, fat, protein and other catabolisms; and from it start out major syntheses of cell compounds and products, and the oxidative phosphorylations that are the cell's main means of energy transfer.

Its discovery came from the study of the respiration of highly aerobic tissues and in particular of preparations from pigeon breast muscle. The minced tissue suspended in well aerated saline rapidly loses its power to consume oxygen, i.e. to respire. The first big advance came in 1934 when Szent-Györgyi, seeking for substances that would enable the minced tissue to maintain its respiration rate, found that succinic, fumaric, malic and oxaloacetic acids were all effective and, moreover, that the additional oxygen taken up was more than equivalent to the acids consumed, i.e. they appeared to be acting catalytically. Szent-Györgyi and his colleagues were able to show that the following interconversions occurred.

34)
$$
\begin{array}{ccccccc}
\text{COOH} & & \text{COOH} & & \text{COOH} & & \text{COOH} \\
| & & | & & | & & | \\
\text{CH}_2 & \xrightleftharpoons{\pm\,2\text{H}} & \text{CH} & \xrightleftharpoons{\pm\,\text{H}_2\text{O}} & \text{CHOH} & \xrightleftharpoons{\pm\,2\text{H}} & \text{C}{=}\text{O} \\
| & & \| & & | & & | \\
\text{CH}_2 & & \text{CH} & & \text{CH}_2 & & \text{CH}_2 \\
| & & | & & | & & | \\
\text{COOH} & & \text{COOH} & & \text{COOH} & & \text{COOH} \\
\text{succinic} & & \text{fumaric} & & \text{malic} & & \text{oxaloacetic}
\end{array}
$$

Shortly after, Krebs and Johnson found that citric acid and α-ketoglutaric acid had similar catalytic activity, and in 1937 they showed that citric acid was formed from oxaloacetic acid by condensation with some compound as yet unknown and could be converted to α-ketoglutaric acid. They suggested that the whole series of reactions was joined to form a catalytic cycle. To support the suggestion, they made use of malonic acid, $HOOCCH_2COOH$, which was known to inhibit succinic dehydrogenase competitively. They showed that succinic acid could be formed from oxaloacetic acid by two pathways,

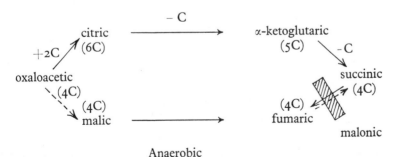

Figure 17 Separation of aerobic and anaerobic formations of succinic acid by malonic acid.

aerobically through citric acid and anaerobically through malic acid (Figure 17).

In such a cycle, inhibition of succinic dehydrogenase with malonic acid would be expected to increase the yield of succinic acid in oxygen and to reduce it in nitrogen; and such indeed was found to be the fact.

More than thirty years later the full enzymology of Krebs' cycle is still open to doubt in detail; but its major features may at present be summarised as in Figure 18.

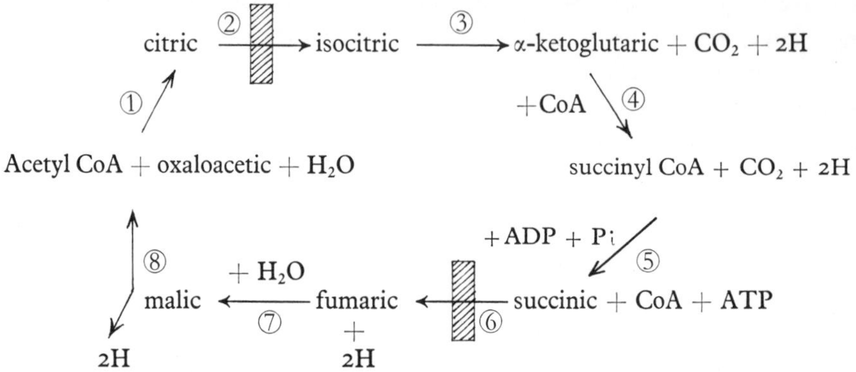

Figure 18 Outline of the citric acid cycle. For enzymes 1 to 8 and inhibitions at 2 and 6 see text.

Reaction 1 is catalysed by 'condensing enzyme' which has been crystallised from pig's heart muscle. It is an aldol condensation with uptake of H_2O

35) $$CH_3COSCoA + \begin{matrix} COOH \\ | \\ C{=}O \\ | \\ CH_2COOH \end{matrix} + H_2O \rightarrow \begin{matrix} CH_2COOH \\ | \\ COHCOOH \\ | \\ CH_2COOH \end{matrix} + CoASH$$

acetyl CoA — oxaloacetic — citric

Reaction 2 is the conversion of citric to isocitric acid by the enzyme aconitase; aconitic acid may or may not be an intermediate. Experiments with deuterium labelling have not supported the implied removal and addition of water.

36) $$\begin{matrix} CH_2COOH \\ | \\ COHCOOH \\ | \\ CH_2COOH \end{matrix} \xrightleftharpoons{aconitase} \begin{matrix} CHOHCOOH \\ | \\ CHCOOH \\ | \\ CH_2COOH \end{matrix} \quad \left[\begin{matrix} CH_2COOH \\ | \\ CCOOH \\ \| \\ CHCOOH \end{matrix}\right]$$

citric — isocitric — aconitic

Aconitase can be specifically inhibited (p. 64) and forms a second point at which the cycle may be interrupted experimentally. When this is done citric acid would be expected to accumulate assuming the cycle to be in operation.

Reaction 3, the conversion of isocitric acid to α-ketoglutaric, is an oxidative decarboxylation with removal of 2H and CO_2. Two separate isocitric dehydrogenases are known, one transferring the 2H to NAD and the other to NADP. The latter is said to form oxalosuccinic acid as an intermediate; but dehydrating and decarboxylating components have not been separated.

37)

$$\begin{array}{llcccc}
\text{CHOHCOOH} & \text{NADP} & \text{NADPH}_2 & \text{COCOOH} & & \text{COCOOH} \\
| & & & | & -CO_2 & | \\
\text{CHCOOH} & \longleftarrow & \longrightarrow & \text{CHCOOH} & \rightleftarrows & \text{CH}_2 \\
| & & & | & +CO_2 & | \\
\text{CH}_2\text{COOH} & & & \text{CH}_2\text{COOH} & & \text{CH}_2\text{COOH} \\
\text{isocitric} & & & \text{oxalosuccinic} & & \text{α-ketoglutaric}
\end{array}$$

The NAD-linked enzyme has not been shown to form oxalosuccinic acid as an intermediate or to catalyse the reverse reaction. On the grounds that it is the more closely connected of the two enzymes with the mitochondria, it is presumed to be the one operative in the citric acid cycle.

Reaction 4, the oxidative decarboxylation of α-ketoglutaric acid, requires the same cofactors as that of pyruvic (α-ketopropionic) acid, and its mechanism is probably similar to that described on p. 56, but resulting in the formation of succinyl-CoA.

38)

$$\begin{array}{llll}
\text{COCOOH} & & \text{COSCoA} & \\
| & & | & \\
\text{CH}_2 & + \text{CoASH} + \text{NAD} \longrightarrow & \text{CH}_2 & + \text{NADH}_2 + CO_2 \\
| & & | & \\
\text{CH}_2\text{COOH} & & \text{CH}_2\text{COOH} & \\
\text{α-ketoglutaric} & & \text{succinyl-CoA} &
\end{array}$$

The soluble enzyme complex has been extracted from animal cells and bacteria and shown to exist in plant mitochondria. Under cellular conditions the sequence is virtually irreversible which makes the cycle also irreversible as a whole.

Reaction 5 is catalysed by the enzyme succinyl-CoA synthetase which has been purified from animal cells and spinach leaves. It is the only reaction of the cycle which directly ('at substrate level') involves the formation of ATP.

39)

$$\begin{array}{llll}
\text{COSCoA} & & \text{COOH} & \\
| & & | & \\
\text{CH}_2 & + \text{ADP} + \text{P}_i \rightleftarrows & \text{CH}_2 & + \text{CoASH} + \text{ATP} \\
| & & | & \\
\text{CH}_2\text{COOH} & & \text{CH}_2\text{COOH} & \\
\text{succinyl-CoA} & & \text{succinic acid} &
\end{array}$$

The enzyme isolated from spinach leaves links the inorganic phosphate directly with ADP; but the animal enzyme is specific for guanosine diphosphate; the resulting guanosine triphosphate transfers its phosphate to the ADP in a reaction catalysed by the appropriate kinase.

Reaction 6, the dehydrogenation of succinic acid, differs from the similar reactions already described in that the 2H are not transferred either to NAD or to NADP, but on account of the more positive potential of succinic acid direct to a flavoprotein, succinic dehydrogenase, (p. 80).

40)
$$
\begin{array}{ccc}
\text{CH}_2\text{COOH} & & \text{CHCOOH} \\
| & \xrightleftharpoons[+2\text{H}]{-2\text{H}} & \| \\
\text{CH}_2\text{COOH} & & \text{HOOCHC} \\
\text{succinic} & & \text{fumaric}
\end{array}
$$

The inhibition of this stage by malonic acid has already been mentioned. As it is competitive it can be 'reversed' by addition of excess succinate; but the affinity ratio is about 50:1 in favour of malonate.

Reaction 7, the hydration of fumaric acid to malic, is catalysed by fumarase

41)
$$
\begin{array}{ccc}
\text{CHCOOH} & & \text{CHOHCOOH} \\
\| & \xrightleftharpoons[-\text{H}_2\text{O}]{+\text{H}_2\text{O}} & | \\
\text{HOOCHC} & & \text{CH}_2\text{COOH}
\end{array}
$$

No cofactors are known to be required.

Reaction 8 is catalysed by malic dehydrogenase which apparently transfers the 2H to NAD. The enzyme prepared from spinach leaves also reduces NADP.

42)
$$
\begin{array}{ccc}
\text{CHOHCOOH} & & \text{COCOOH} \\
| \quad +\text{NAD} & \rightleftharpoons & | \quad +\text{NADH}_2 \\
\text{CH}_2\text{COOH} & & \text{CH}_2\text{COOH} \\
\text{malic} & & \text{oxaloacetic}
\end{array}
$$

The oxaloacetic acid so formed completes the cycle re-entering reaction 1.

The nature of the evidence

It now seems probable that the citric acid cycle occurs in almost all cells. To demonstrate the existence of any metabolic reaction sequence is difficult; there is no universal criterion or group of tests, and the general discussion is outside our scope. It would not be so bad if metabolism consisted of a few straightforward and mutually exclusive sets of reactions—as was at one time supposed. It is now clear that it includes not only linear, but also branching, cyclic and spiral sequences which interconnect at frequent points and which may shunt, reverse and supplement one another. Also, the metabolic equipment of a given cell is not a constant quantity; the appearance and disappearance of adaptive enzymes, once thought to be a speciality of the bacteria, is actually much more widely spread.

It will be convenient to take the citric acid cycle as an example of the sort of evidence that can be brought forward, both because of the importance of the cycle and because of the great amount of effort expended on it. The main facts in its favour fall under five main headings.

Intermediates
The most obvious question is whether the proposed intermediates occur. The answer is that they have all been found in a wide range of plant, animal and microbial cells, though of course, in very varying amounts. Oxaloacetic acid is usually present in only very low concentrations whereas malic may be present in quite high ones. This is not solely a matter of the steady state equilibria involved in the cycle reactions. Malic acid is formed from other sources, and some of the other acids may be also.

Enzymes
All the postulated enzymes have been identified in numerous types of cell, and some of them obtained in a high state of purity. It would, of course, be helpful to be able to show that they are not only present, but are there in sufficient amounts to account for the observed rates of respiration of the cell or tissue. Unfortunately, this may involve great technical difficulties—strictly speaking the quantitative extraction of pure enzymes, something which has never yet been achieved. Partially purified systems may be inhibited or accelerated by impurities still present. It is, for example, difficult to demonstrate succinic dehydrogenase activity in many plant tissue extracts, because the cells contain malonic acid. Because of the slow rate at which yeast preparations oxidise many of the cycle intermediates, it was supposed for a time that the cycle did not constitute a respiratory pathway in yeast. Conversely, it is evident that a high rate of activity in an extracted system, as for example in the oxidations due to polyphenolases, is no guarantee of a similar rate in the cell.

Addition of intermediates
The oxygen consumption of numerous types of cell is accelerated by the addition of, for example, succinic, fumaric and α-ketoglutaric acids under suitable conditions. With microorganisms and tissues like plant roots the intermediates can be fed directly; with others like muscle and liver such feeding

necessitates either slicing or mincing, and even then the substance tested may be unable to penetrate the cell surface; this often happens with citric acid. Conversely, a fed substance may give rise to increased oxygen demand even though it is not respired (cf., alcohol p. 52).

^{14}C-labelling

If ^{14}C-labelled pyruvate or acetate is supplied, ^{14}C-labelling quickly appears on the intermediates in the cycle, and on those amino acids derived directly from it, i.e. on glutamic acid from α-ketoglutaric and on aspartic from oxalo-acetic; $^{14}CO_2$ is given off.

Inhibitors

The effect of malonic acid in reducing oxygen consumption and causing succinic acid to accumulate due to its inhibition of succinic dehydrogenase has already been mentioned. Respiration can be restored by addition of fumaric acid 'beyond the block', or by excess succinic. Added fluoroacetic acid is converted in tissues to fluorocitric acid which inhibits aconitase and citric acid accumulates in the tissues. Aconitase requires loosely held Fe^{2+} as cofactor, possibly to form an enzyme—Fe^{2+}—citric acid carbonium ion complex. Fe^{2+} chelating agents, which do not remove the strongly bound iron in haems, compete with citric acid for the aconitase Fe^{2+}. Dilute 2,2'-bipyridyl reduces the oxygen consumption of barley roots and embryos and citric acid accumulates in the tissues. Excess citric acid reverses the inhibition. It is also interesting that oxygen pressures above 1 atm, that may be supposed to oxidise the Fe^{2+} in the tissues, also cause an accumulation of citric acid in apples and potatoes and at the same time reduce the amount of α-ketoglutarate present.

It must be noted that failure of an inhibitor to inhibit oxygen consumption or carbon dioxide output does not necessarily rule out the affected reaction as part of the respiration pathway, even when it is certain that the inhibitor has entered the cell. Suppression of the normal path may have led to the increased operation of a normally minor shunt.

Such points as the identification of intermediates and enzymes are evidence of the cell's capabilities, rather than of what it is actually doing at any particular moment. It depends on the point of view how important a distinction this is. To the chemical engineer seeking to obtain a good yield of a difficult biochemical for medical or technical purposes, what the cell can be controlled into doing is more important than what it does left to itself. The cell physiologist who wants to understand the natural order of events sets himself in some ways a more difficult task. Indeed, it is sometimes said to be an impossible one, because every experiment involves some change in the natural state of affairs. But this overlooks the fact that the 'natural state' itself varies through fairly wide ranges; cells do not normally live in electronically controlled growth cabinets.

Enough has perhaps been said to make it clear that there are no absolute criteria giving forthright answers to metabolic questions. This is why the investigation of its pathways has not proceeded by successive logical steps; but has progressed only at the expense of frequent rethinking and doubling back as well as by the inspired leaps forward.

The formation of acetyl CoA from fatty acids

So far, it has been necessary only to consider carbohydrates as the main source of respired materials. But many tissue cells, particularly those of liver and the storage tissues of fatty seeds, utilise fats without first converting them to carbohydrates. The fats are hydrolysed to long chain fatty acids and glycerol. The fate of the glycerol is considered on p. 128. The fatty acid chains are shortened by β-oxidation, i.e. by the stepwise cutting off of 2C fractions at the carboxyl end. This happens as follows. The fatty acid first reacts with CoA in the presence of ATP

43) $RCH_2CH_2COOH + CoASH + ATP \xrightarrow{\textit{thiokinase}} RCH_2CH_2COSCoA +$
$AMP + PP_i$

    fatty acid   acyl CoA

The inorganic pyrophosphate (PP_i) is hydrolysed and removed; any accumulation slows down the reaction. The acyl CoA is dehydrogenated with FAD as receiver, forming an unsaturated acyl CoA

44) $RCH_2CH_2COSCoA + FAD \xrightarrow{\textit{acyl CoA dh}} RCH=CHCOSCoA + FADH_2$

This resembles the unsaturation of succinic acid to fumaric in that the methylene hydrogen atoms do not have a sufficiently negative redox potential to reduce NAD and are transferred to the more positive FAD instead. Again as in fumaric acid, the double bond is hydrated giving, instead of malic acid, a hydroxyacyl CoA,

45) $RCH=CHCOSCoA + H_2O \xrightarrow{\textit{enoyl hydrase}} RCHOHCH_2COSCoA$

which is in turn oxidised to the corresponding ketoacyl CoA, NAD now being the hydrogen acceptor just as during the oxidation of malic to oxaloacetic acid

46) $RCHOHCH_2COSCoA + NAD \xrightarrow{\textit{β-hydroxyacyl CoA dh}} RCOCH_2COSCoA +$
$NADH_2$

As a final step, a second molecule of CoA reacts with the ketoacyl CoA, the products being acetyl CoA and an acyl CoA with two carbon atoms fewer than the original fatty acid. This shortened acyl CoA can then in its turn pass through the β-oxidation sequence (Figure 19). During each shortening two pairs of hydrogens are transferred to the oxidation chain, one pair via an FAD-linked dehydrogenase and the other through an NAD-linked one. These are in addition to those to be derived via the acetyl CoA.

In some circumstances proteins may also make a contribution to the respired materials. Some amino acids derived from their hydrolysis give rise to acyl CoA compounds; e.g. leucine to acetyl CoA + propionyl CoA, isoleucine to acetyl CoA + acetoacetic acid.

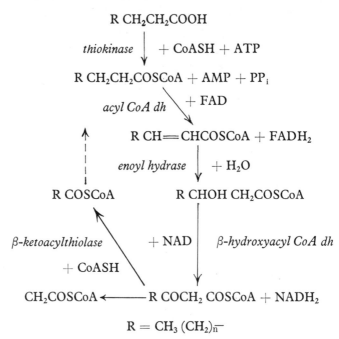

$$R = CH_3 (CH_2)_{\overline{n}}$$

Figure 19 The β-oxidation of fatty acids. One turn of the spiral is shown reducing the chain length by 2 carbons. Enzymes are shown in italics.

The glyoxylic acid cycle

Some bacteria, yeast and moulds can survive and grow when acetic acid is the only source of carbon. They possess two enzymes, isocitritase and malate synthetase, which together make possible a shunt of the citric acid cycle. Isocitritase catalyses the reaction

47)

$$\underset{\text{isocitric}}{\begin{array}{l} \text{CHOHCOOH} \\ | \\ \text{CHCOOH} \\ | \\ \text{CH}_2\text{COOH} \end{array}} \quad \rightleftharpoons \quad \underset{\text{glyoxylic}}{\text{CHOCOOH}} + \underset{\text{succinic}}{\begin{array}{l} \text{CH}_2\text{COOH} \\ | \\ \text{CH}_2\text{COOH} \end{array}}$$

and malate synthetase causes the glyoxylic acid to condense with acetyl CoA to give malic acid

48)
$$CHOCOOH + CH_3COSCoA + H_2O \rightleftharpoons \begin{matrix} CHOHCOOH \\ | \\ CH_2COOH \end{matrix} + CoASH$$

    glyoxylic   acetyl CoA   malic   CoA

Assuming the acetic acid to be 'activated' as usual to acetyl CoA, the cycle can be summarised as in Figure 20.

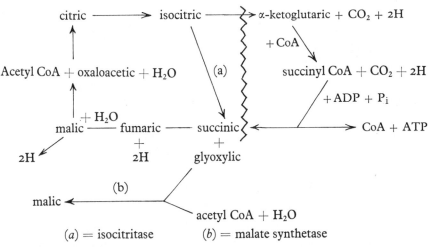

(a) = isocitritase (b) = malate synthetase

Figure 20 The glyoxylic acid cycle. Reactions shown in small type to the right of the zig-zag line are those of the citric acid cycle which the cycle short circuits.

The cycle apparently only occurs when the organisms are supplied with acetate in place of carbohydrate. It does not happen in animals and the only tissue cells in which it is known are those of fatty seeds. During germination these produce large quantities of acetyl CoA by the β-oxidation of their fatty acids. The interesting thing is that they also produce isocitritase and malate synthetase at the same time, and only while the fat-breakdown is going on. They afford a well-documented example of adaptive enzyme formation in an advanced cell type.

Although the citric acid and glyoxylic acid cycles share at least seven common reactants their outcomes are quite different. The decarboxylating stages of the citric acid cycle (on the right side of Figure 20) are avoided in the glyoxylic cycle. Although the citric acid cycle is the point of departure of several well-known lines of synthesis, such as amino acids from its α-keto acids and porphyrins from succinyl CoA, each turn of the cycle involves a loss of 2 CO_2 and no net gain of carbon. To the extent that any of the intermediates are drawn off into syntheses, the cycle is diminished. Each turn of the glyoxylic cycle on the other hand results in the formation of an extra malic acid molecule, which, however, rarely accumulates as such. It may be fed back into the cycle to replace other acids drawn off in syntheses; in fatty seeds it is largely synthesised to carbohydrate. The cycle thus appears to be a means by which certain cell types turn to account an unusually abundant formation of acetyl CoA. At present, at least, it is not known to occur in a great variety of cells.

The pentose phosphate pathway

The oxidative decarboxylation of glucose-6-phosphate on the other hand occurs in a wide variety of cell types from liver, lens and red blood cells to moulds such as *Penicillium chrysogenum*, aerated yeasts and bacteria including *E. coli*. It also occurs in many plant tissues. It may be regarded as an aerobic development of the anaerobic process found in *Leuconostoc* (p. 47); but, whereas the anaerobic version appears to be rare, the aerobic one is common. The first two steps, the oxidation of glucose-6-phosphate to 6-phosphogluconate and the subsequent oxidation of the latter to ribulose-5-phosphate and CO_2 are similar in the two versions. The aerobic glucose-6-phosphate dehydrogenase and 6-phosphogluconic dehydrogenase respectively, are both specific to NADP and both transfer hydrogen to the B side of its nicotinamide ring. The sum of these two oxidising and decarboxylating steps is

49) $CH_2O(P)(CHOH)_4CHO + H_2O + 2\,NADP \rightarrow CH_2O(P)(CHOH)_2COCH_2OH$
$$+ CO_2 + 2NADPH_2$$

glucose-6-phosphate   ribulose-5-phosphate

The fate of the ribulose-5-phosphate is interesting (Figure 21). It is acted on directly by two enzymes; an isomerase that converts it from the keto to the corresponding aldo form and an epimerase that changes its configuration about carbon atom 3.

$$
\begin{array}{ccccc}
\text{CHO} & & \text{CH}_2\text{OH} & & \text{CH}_2\text{OH} \\
| & & | & & | \\
\text{HCOH} & & \text{C=O} & & \text{C=O} \\
| & \xleftarrow{\ isomerase\ } & | & \xleftarrow{\ epimerase\ } & | \\
\text{HCOH} & \xrightarrow{\hspace{1cm}} & \text{HCOH} & \xrightarrow{\hspace{1cm}} & \text{HOCH} \\
| & & | & & | \\
\text{HCOH} & & \text{HCOH} & & \text{HCOH} \\
| & & | & & | \\
\text{CH}_2\text{O}(P) & & \text{CH}_2\text{O}(P) & & \text{CH}_2\text{O}(P) \\
\text{D-ribose-5-phosphate} & & \text{D-ribulose-5-phosphate} & & \text{D-xylulose-5-phosphate}
\end{array}
$$

50)

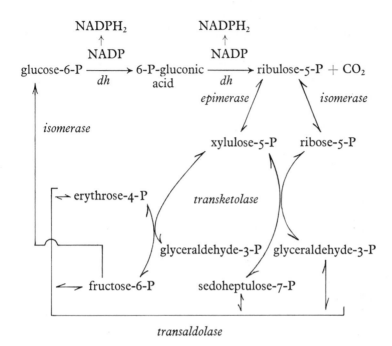

Figure 21 Outline diagram of the pentose phosphate pathway. Enzymes are shown in italics.

Transketolase transfers ketol, CH_2OHCO-, groups to aldehydes from donors with the configuration

$$
\begin{array}{c}
CH_2OH \\
| \\
C{=}O \\
| \\
HOC{-}H \\
|
\end{array}
$$

for example, from xylulose-5-phosphate to ribose-5-phosphate, the products being sedoheptulose-7-phosphate and glyceraldehyde-3-phosphate

xylulose-5-phosphate   glyceraldehyde-3-phosphate

51)

$$
\left.
\begin{array}{c}
CH_2O\textcircled{P}(CHOH)_2COCH_2OH \\
+ \\
CH_2O\textcircled{P}(CHOH)_3CHO
\end{array}
\right\}
\underset{\textit{transketolase}}{\rightleftharpoons}
\left\{
\begin{array}{c}
CH_2O\textcircled{P}CHOHCHO \\
+ \\
CH_2O\textcircled{P}(CHOH)_4COCH_2OH
\end{array}
\right.
$$

ribose-5-phosphate   sedoheptulose-7-phosphate

Transaldolase, transfers aldol groups, $CH_2OHCOCHOH-$, for example from the sedoheptulose-7-phosphate to the glyceraldehyde-3-phosphate, both formed by the transketolation above. The products are then further fructose-6-phosphate and erythrose-4-phosphate.

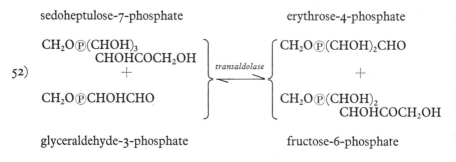

             sedoheptulose-7-phosphate   erythrose-4-phosphate

52)

             glyceraldehyde-3-phosphate   fructose-6-phosphate

Erythrose-4-phosphate is an alternative acceptor of CH_2OHCO- from xylulose-5-phosphate, the products being fructose-6-phosphate and glyceraldehyde-3-phosphate.

             xylulose-5-phosphate   glyceraldehyde-3-phosphate

53)

             erythrose-4-phosphate   fructose-6-phosphate

Phosphohexoisomerase converts the fructose-6-phosphate to glucose-6-phosphate the starting point of the cycle.

All the above reactions, except the initial dehydrogenation of glucose-6-phosphate are reversible, the enzymes are readily soluble and transketolase and transaldolase have rather low specificity; other reactions are possible. Since glucose-6-phosphate is reformed the reactions may be regarded as setting up a cycle; but, owing to the loss of CO_2 in the decarboxylation of 6-phosphogluconate, only 5 molecules of glucose-6-phosphate can be reconstituted for each 6 utilised. Recycling may lead eventually to the complete oxidation of each glucose-equivalent to 6 CO_2. The system has therefore been regarded as an alternative respiratory pathway to that via glycolysis and the citric acid cycle. Since glyceraldehyde-3-phosphate is an end product it may, in fact, lead back into glycolysis, and for this reason is sometimes called the hexosemonophosphate shunt.

Like the glycolysis–citric acid systems the pentose phosphate pathway cannot in any case be regarded solely as an energy transformer. It reversibly produces a pool of C_3, C_4, C_5, C_6 and C_7 sugar phosphates, from which any one of them may be drawn off. Perhaps most interesting of all is its formation of ribose phosphate which plays so large a part in the synthesis of nucleotides and nucleic acids. It is worth noting that, owing to the reversibility of the later reactions of the cycle, the ribose phosphate could be formed anaerobically without involving any oxidation stage. Erythrose is a precursor of some aromatics, such as dehydroquinic acid.

Unlike the oxidations of the glycolysis-TCA system those of the pentose phosphate pathway require NADP. Under aerobic conditions the $NADPH_2$ so generated does not reduce glycolysis intermediates like the PPP of *Leuconostoc*; but neither does it appear to be commonly oxidised through systems leading directly to oxygen (cf., p. 129). It might, in fact, be regarded as a major function of the PPP to produce 'reducing power' for synthetic and other purposes in a form that cannot be siphoned off through the electron transport chain.

$NADPH_2$ is the reductant required for reductive carboxylation of pyruvic acid (p. 73), an alternative method of malic acid formation. It is also required in the synthesis of fats in adipose and mammary tissues where the pathway is said to be particularly active. Administration of insulin, which stimulates fatty acid synthesis, is also said to increase the metabolic flow through the pentose phosphate pathway, while decreasing that through the citric acid cycle. In red cells a minimal concentration of reduced glutathione is required to maintain the normal behaviour of the cell membrane and prevent it breaking down. The glutathione is kept reduced by $NADPH_2$, which in turn is kept reduced by reactions of the pentose phosphate pathway, i.e. by oxidation of glucose-6-phosphate. In red cells deficient in glucose-6-phosphate dehydrogenase, the ratio $NADPH_2/NADP$ decreases and the cells may lyse. This is most likely to happen in older cells, which have no nucleus, and are unable to resynthesise lost enzymes.

Both glycolysis-TCA and the pentose phosphate complex are now known to occur in many differing types of cells. Frequently they have been shown to occur simultaneously, and efforts have been made, especially by means of ^{14}C-labelling experiments, to assess the relative magnitudes of the two flows. In view of the complexities of the two processes themselves and of their many connections with one another and with multiple syntheses, it is perhaps not surprising that the results have so far led to discussion rather than determination.

Exergonic carboxylations

Phosphoenolpyruvate (PEP) is a high-energy phosphate (p. 38) and in glycolysis its readily available energy is used to form ATP.

54) $CH_2{=}CO{\sim}\textcircled{P}COOH + ADP \xrightleftharpoons{\textit{pyruvic kinase}} CH_3COCOOH + ATP$

Under cell conditions ΔG is probably about -6 kcal; and the reaction is not readily reversible.

In the presence of the widely distributed enzyme PEP carboxylase the available energy can be used to drive a carboxylation, i.e. a 'fixation of carbon dioxide,' the product being oxaloacetic acid.

55) $CH_2{=}CO{\sim}\textcircled{P}COOH + CO_2 + H_2O \xrightleftharpoons{\textit{PEP carboxylase}} HOOCCH_2COCOOH + HO\textcircled{P}$

The cellular ΔG is again about -6 kcal; equilibrium lies well to the right and the reaction is not readily reversed.

In the cells of succulent plants of the family Crassulaceae this system is particularly active and leads via reduction by malic dehydrogenase to the accumulation of large amounts of malic acid in the cell vacuoles. Labelling experiments suggest, however, that it cannot be the only route.

A third reaction, catalysed by PEP carboxykinase, effects phosphate transfer and carboxylation simultaneously.

56) $CH_2{=}CO{\sim}\textcircled{P}COOH + ADP + CO_2 \xrightleftharpoons{\textit{PEP carboxykinase}} HOOCCH_2COCOOH + ATP$

In this reaction the available energy is fully utilised and the reaction is readily reversed towards PEP formation. It is probably the route by which oxaloacetate from the excess malate arising during the germination of fatty seeds is resynthesised to carbohydrates.

Less active substances than PEP may be carboxylated under suitable conditions. Liver cells contain a pyruvic carboxylase which catalyses the carboxylation of (keto) pyruvic acid utilising the energy of ATP.

57) $CH_3COCOOH + CO_2 + ATP \xrightleftharpoons{\textit{pyruvic carboxylase}} HOOCCH_2COCOOH + ADP$

     pyruvic   oxaloacetic

The oxaloacetic acid may be converted to PEP by a PEP carboxykinase which differs from the plant enzyme mentioned (Eqn 56) in using guanidine triphosphate in place of ATP. Liver cells resynthesise glycogen from the lactic acid received via the blood stream from fatigued muscle. These reactions probably represent the two initial steps, after which glycolysis is readily reversible (Figure 22).

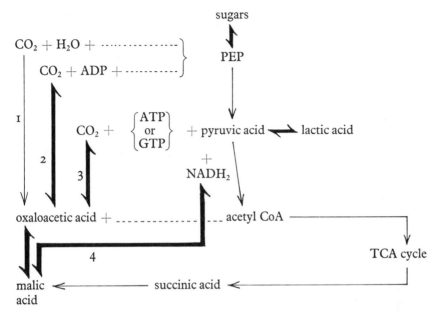

Figure 22 Exergonic carboxylations and some associated reactions. Enzymes: 1 PEP carboxylase, 2 PEP carboxykinase, 3 pyruvic carboxylase, 4 malic enzyme. Thin arrows denote reactions which are effectively irreversible in the cell; heavy arrows those that are reversible.

Carboxylation of pyruvic acid also occurs in the presence of $NADPH_2$. In the reaction

58) $CH_3COCOOH + CO_2 + NADPH_2 \xrightleftharpoons{\textit{malic enzyme}} HOOCCH_2CHOHCOOH + NADP$

   pyruvic   malic

ΔG_0 is about -0.3 kcal. If the $NADPH_2$ is continuously regenerated there is continuous formation of malic acid from pyruvic.

Ribulose-1,5-diphosphate is carboxylated to 2 triose molecules by ribulose diphosphate carboxylase. The enzyme is present in chloroplasts in large amounts and also in chemosynthetic cells, though probably absent from heterotrophs. The reaction is more or less irreversible around neutral pH and is regarded as the initial dark fixation of CO_2 in photosynthesis.

It will be noted that these carboxylations are all to some extent exergonic and, although they may represent a gain of carbon to the cell, they involve a direct loss of energy. This is provided by some preformed energy-rich intermediate such as PEP or by preformed ATP, i.e. by energy deriving ultimately from carbohydrate. The carboxylation of ribulose-1,5-diphosphate is exceptional only in that its energy bill is ultimately defrayed by the light reaction.

The decarboxylations, recarboxylations and all the associated reactions considered in this chapter are almost all easily reversible. That is, they do not involve any considerable loss or transfer of free energy. What they evidently do afford the cell is a variety of interconnections between the manifold aspects of metabolism; between the almost infinitely varied products of cell synthesis and the energy-providing reactions themselves. They occur widely in many different kinds of cells and illustrate the versatility of metabolism. They may perhaps be regarded as economising the material resources of cells as ATP-formation conserves their energy.

Further Reading

Bartley, W., Birt, L. M. and Banks, P. (1968) *The Biochemistry of the Tissues* Chapters 4 & 8. Wiley, London and New York.

Dagley, S. and Nicholson, D. E. (1970) *An Introduction to Metabolic Pathways.* Blackwell Scientific Publications, Oxford.

Kornberg, H. R. and Beevers, H. (1957) The glyoxalate cycle as a stage in the conversion of fat to carbohydrate in castor beans. *Biochem. Biophys. Acta*, **26,** 531–7.

Kornberg, H. R. and Krebs, H. A. (1957) Synthesis of cell constituants from C_2 units by a modified tricarboxylic acid cycle. *Nature*, **179,** 988.

Krebs, H. A. and Lowenstein, J. M. (1960) The tricarboxylic acid cycle. In *Metabolic Pathways*, vol. 1. Academic Press, New York.

Lowenstein, J. M. (1969) *The Citric Acid Cycle*. Dekker, New York.

Neil, M. W. (1965) *Vertebrate Biochemistry*. 2nd ed. (Chapter 10). Pitman, London.

Walker, D. A. (1966) Carboxylation in plants. *Endeavour*, **25,** 21–6.

5 *Oxidation*

THE reaction paths described in the last chapter are aerobic, because they continue only so long as the $NADH_2$ and reduced succinic dehydrogenase they give rise to are continuously reoxidised at the expense of dissolved oxygen. The reaction

59)
$$NADH_2 + \tfrac{1}{2}O_2 \longrightarrow NAD + H_2O$$

has ΔG_0 of approximately 52 kcal and, needless to say, does not occur in cells in a single step; at a moderate estimate there are always at least half a dozen successive hydrogen transfers.

Oxidation-reduction potentials

Cells present us with a bewildering choice of candidates for the job of 'hydrogen transporter', i.e. of readily reducible and oxidisable substances. The readiness with which a substance loses electrons, i.e. its status as a redox body, can be expressed as an EMF, and, where the electrons will pass to an electrode any two such tendencies can be compared in a suitable electrolytic cell. The system in which molecular hydrogen is the reductant and the hydrogen ion is the oxidant

60)
$$\tfrac{1}{2}H_2 \rightleftharpoons H^+ + e$$

is taken as standard, with hydrogen pressure $= 1$ atm and hydrion concentration unity (pH $= 0$). In the half cell round the other electrode the concentration of reductant and oxidant e.g. $Fe^{2+} \rightleftharpoons Fe^{3+}$, may be varied as desired. With a liquid junction between the two half cells and connection via a potentiometer between the electrodes, an EMF is recorded which follows a sigmoid curve becoming more positive with increasing oxidation. The potential recorded at 25°C and pH 0 with reduced and oxidised forms present in equal concentrations is the standard redox potential, E_0. For biological purposes measurements have mostly been made at pH 7 to give a standard E'_0.

In aqueous solutions the removal of an electron from the field of a hydrogen atom requires relatively little energy; but a much greater amount would be required to remove it from an oxygen atom. Consequently an electron moving from hydrogen to oxygen will yield energy as if passing from the negative to the positive pole of an electric field. Similarly, in making the partial moves between intermediate carriers of successively increasing positive (or decreasing negative) potential it will be decreasing free energy. The standard redox potential of the couple, $NADH_2 \rightleftharpoons NAD$, is -0.32 V and in $H_2O \rightleftharpoons \frac{1}{2} O_2 + H_2$ it is $+0.82$ V. Transporters of hydrogen or electrons between $NADH_2$ and O_2 may be expected to have E'_0 values between these two limits. The total drop in free energy is approximately 52 kcal. The drop at any particular step may be calculated from

61) $$\Delta G = -n \text{ F } \Delta E'_0$$

Where F is 23 068, the Faraday constant in cal per volt equivalent, and n the number of electrons involved.

Though in general, movement of electrons may be expected from one redox pair to another with a more positive E'_0, it must be remembered that the actual EMF varies with temperature, pH and the concentration ratio of oxidised to reduced partner within the pair. Especially, therefore, where the difference of potential between redox pairs is small, implying a small change of free energy, the direction of electron flow might become the reverse of that suggested by the standard potential.

The respiratory oxidation chain

Oxidation chains are not identical in all cells; nor is it clear that any given cell has only one. There is, however, one series now known to occur very widely and to be likely to account for at least a major part of the respiration when it is present. It is perhaps as 'normal' in the final stages of cell respiration as glycolysis is in the initial; but probably shows more variation within its own limits than glycolysis does.

In the 1920s it was already known from the work of Wieland that metabolic substrates are oxidised by the removal of hydrogen by the enzymes we now know as dehydrogenases. On the other hand Warburg was emphasising that enzyme-bound Fe^{2+} was the agent through which cells most often react with molecular oxygen. He admitted the possibility of an occasional similar action in Cu^+. As early as 1885 MacMunn had discovered spectroscopically that a wide range of tissues 'from echinoderms to man' possess haems; iron-containing compounds, which in the reduced state have characteristic visible absorption spectra that disappear when the haems are oxidised. This behaviour could be observed particularly well in the wings of insects. MacMunn proposed a respiratory function for his haems, but he was argued down. In 1925 Keilin rediscovered MacMunn's 'histohaematins', in a wide range of cells including animal cells, yeast and those of many non-green plant tissues. He rechristened them cytochromes and, from the different rates at which the bands appeared and disappeared, decided that three different pigments must be present, which he named cytochromes *a*, *b* and *c*. It was eventually realised that they were carriers between Wieland's dehydrogenases and Warburg's iron-containing 'respiratory enzyme' which reacted directly with oxygen.

Subsequent research has shown the cytochromes to be so widespread, and to conform so closely in their behaviour to what would be expected, that the oxidation chain containing them has come to be generally regarded as the normal one in respiration. It has also been shown, however, that there are many more cytochromes than three and that other carriers also participate in the chain.

In the present state of knowledge, the oxidation of $NADH_2$ may be outlined as in Figure 23. Hydrogen or electrons pass from $NADH_2$ to a flavoprotein and thence by way of ubiquinone (CoQ) through a series of cytochromes to oxygen.

There are two main reasons for placing the carriers in the order indicated. The first is that this is the order suggested by their standard redox potentials, E_0' (with the exception of CoQ, which is insoluble in water). This as it stands

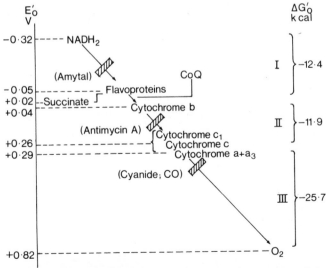

Figure 23 The respiratory oxidation chain, showing the redox potentials of the principal components (left) and the major decreases of free energy involved (right).

is not a strong argument, because the actual potentials operating in the cell depend on the relative proportions of each carrier in the oxidised and reduced states. The difference of E_0' between cytochrome c and cytochrome a, for example, is very small; and, if cytochrome c Fe^{2+}/cytochrome c Fe^{3+} were small and the corresponding ratio for cytochrome a were high, an electron flow would be expected from a to c and not in the direction shown. It has, however, been demonstrated with a heart muscle preparation oxidising succinic acid that the ratios of the oxidised/reduced forms of the cytochromes b, c and a do not depart widely from 1. When, therefore, the actual potential is calculated from

$$62) \qquad\qquad E = E_0' + \frac{0.06}{n} \ln \frac{[\text{ox}]}{[\text{red}]}$$

the second term on the right virtually vanishes and $E = E_0'$ (Table 4).

TABLE 4

Oxidation-reduction potentials (E) of cytochromes in a heart muscle preparation oxidising succinic acid; from an experiment by B. Chance.

Cytochrome	E_0'	$\dfrac{[\text{ox}]}{[\text{red}]}$	E
b	0.0	40/60	− 0.01
c	+ 0.26	49/51	+ 0.26
a	+ 0.29	59/41	+ 0.30

E_0' = potential of isolated system

The value of $E_0' = 0.0$ for cytochrome b is the value observed in the presence of structural protein. Highly purified cytochrome b (e.g. from *E. coli*) gives an E_0' value of −0.34. A similar shift occurs with the flavoproteins, which usually show E_0' between 0.0 and −0.05. Riboflavin, the parent compound of the flavins, when dissociated from protein has $E_0' = -0.22$.

The second line of evidence, that confirms the thermodynamic suggestion, is derived from experiments with respiratory inhibitors. As shown in Figure 23 there are three points at which the chain can be interrupted by specific poisons. When aerobic respiration is proceeding at a steady rate with a continuous supply of substrate to keep the $NADH_2$ reduced, the intermediate carriers are in varying states of oxidation. The a_3 component of cytochrome $a + a_3$ is inhibited by dilute cyanide and when this is added all the components of the chain go into the fully reduced state. This can be observed spectroscopically, since all members show characteristic changes according to their redox state. The addition of a strong reducer such as dithionite does not strengthen the bands or cause any new bands to appear (if the experiment is being done with extracted mitochondria) so it may be presumed that the natural electron donors cause complete reduction of all electron carrier molecules in mitochondria.

Application of antimycin A or BAL similarly cause $NADH_2$, flavoprotein and cytochrome b to become reduced; but the cytochromes c and $a + a_3$ become fully oxidised. Amytal (sodium amobarbital) causes only the $NADH_2$ to become fully reduced. Providing that one is dealing with a simple chain of transfers, these results are in accordance with the order suggested. They do not prove, however, that the carriers are a continuous chain; the same result might still be observed, if one or more were on a side branch; and this has in fact been suggested for cytochrome b.

The carriers included in Figure 23 fall into three main classes.

Flavoproteins

The yellow enzymes have firmly attached prosthetic groups which always include the base dimethyl isoalloxazine, which is capable of oxidation and reduction as follows

63)

The first of these enzymes, the 'old yellow enzyme', was isolated by Warburg and Christian. It catalysed the oxidation of glucose-6-phosphate in red blood cells, probably by reoxidising $NADPH_2$ formed as follows

64) $CH_2O\textcircled{P}(CHOH)_4CHO + NADP + H_2O \rightleftharpoons CH_2O\textcircled{P}(CHOH)_4COOH + NADPH_2$

The hydrogen was transferred directly to oxygen with formation of hydrogen peroxide

65) $NADPH_2 + O_2 \xrightarrow{\text{old yellow enzyme}} NADP + H_2O_2$

The flavoproteins can mostly react with oxygen, but usually only sluggishly. The flavoprotein that acts as $NADH_2$ dehydrogenase has been isolated from heart muscle and from mitochondria. Its prosthetic group is probably flavin mononucleotide (FMN) in which the isoalloxazine base is linked with ribityl-phosphate,—$CH_2(CHOH)_3CH_2O\,®$, at R in Eqn 63. Although called a nucleotide, FMN includes the pentahydric alcohol, ribitol, not the sugar ribose. Without the phosphate group it is riboflavin (a vitamin of the B group) and is the part responsible for the yellow colour. It is abundant in egg yolk. $NADH_2$ dehydrogenase contains 1 FMN group per molecule of enzyme (MW about 1 000 000) and 2–4 atoms of non-haem iron, which probably also take part in the carrier function, since they undergo oxidation and reduction during the catalysis. Outside the cell, the enzyme may behave directly as a cytochrome *c* reductase and is not affected by antimycin A.

Succinic dehydrogenase is also a flavoprotein and the oxidation of succinic acid does not proceed through NAD, its redox potential being too positive. Succinic dehydrogenase as isolated from ox-heart mitochondria has 1 mole-

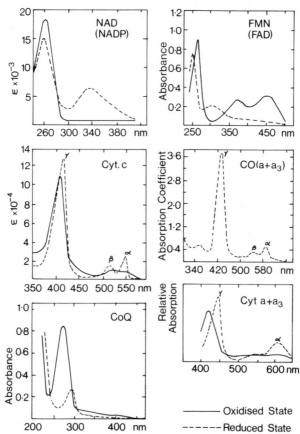

Figure 24 Absorption spectra of some principal components of the respiratory oxidation chain and the action spectrum of the carbonyl compound of the terminal oxidase, CO (a + a₃).

cule of FAD and 4 atoms of non-haem iron per molecule of enzyme. FAD is flavin adenine dinucleotide with the structure.

flavin—ribityl—phosphate
|
adenine—ribose—phosphate

The extracted enzyme does not reduce cytochrome *c* directly. Extracts contaminated with cytochromes *b* and *c*, do so.

The absorption spectrum of FMN is shown in Figure 24. It will be noticed that it is the oxidised form that has the conspicuous absorption bands at 375 nm and 450 nm. The spectrum of FAD is similar and those of the corresponding flavoproteins differ only slightly.

The cytochromes

Cytochromes are all proteins of relatively low molecular weight with tightly bound haems as prosthetic groups. The proteins, the haems and the methods of their binding all differ in different cytochromes. The basic configuration of a haem is porphin, four pyrrole rings united by —CH= groups. In the haems an atom of iron is placed centrally, co-ordinately linked with the nitrogens of the pyrrole rings. The different haems (iron porphyrins) are distinguished by the varying side chains attached to the available pyrrole carbons. The molecule is disc shaped and it is enmeshed within the foldings of the protein peptide helices.

On account of its readier isolation cytochrome *c* has been more fully studied than the others. Its prosthetic group is iron protoporphyrin IX identical with the haem of haemoglobin. It is a resonance hybrid like all porphyrins and has the structure indicated in Figure 25. It is linked to its protein by two thio-ether bonds at the edge of the porphyrin disc, and by two co-ordinate bonds with the iron, which project at right angles above and below the porphyrin as shown in Figure 25. The proteins of cytochrome *c* as isolated from a variety of tissues have been shown to be very similar; but that from *Pseudomonas aeroginosa* had a widely different amino acid complement. The characteristics of the main cytochromes believed to be associated with respiration are summarised in Table 5. There are others associated with photosynthetic electron transfer in chloroplasts and some of unknown function.

All the cytochromes have absorption spectra of a type characteristic of haemochromogens. Three well marked peaks are shown in the reduced Fe^{2+} state; the α and β peaks in the visible wave lengths flatten out on oxidation and the γ peak below the visual range shifts its position. The absorption spectra of pure cytochrome *c* are shown in Figure 24. A third curve might show the difference, reduced-oxidised, spectrum. This is the spectrum obtained when a

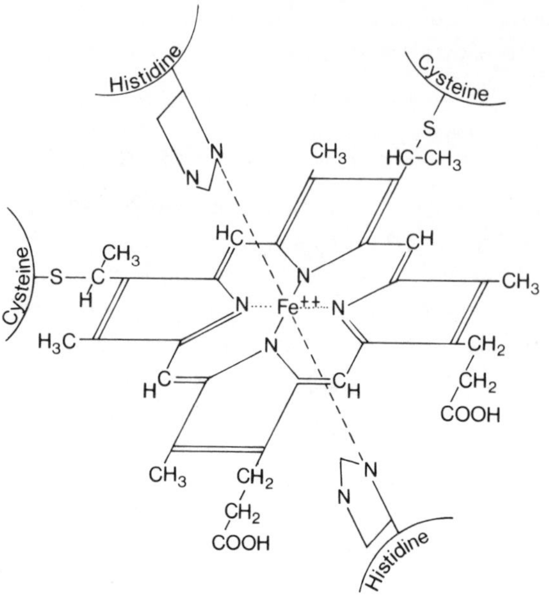

Figure 25 Ferrous protoporphyrin IX (haem) as linked with the protein in cytochrome *c*.

solution of the fully oxidised substance is used as reference in place of the pure solvent. In examining live tissues or tissue extracts spectroscopically, considerable difficulties arise from unavoidable light scattering and background absorption due to other substances. These difficulties are minimised if a fully oxidised sample is used as control. Much of the work on cytochrome behaviour in cells has been done in terms of reduced-oxidised difference spectra.

Cytochrome oxidase

It will be seen from Table 5 that the reactions of cytochrome a_3 differ from those of the other cytochromes. Although the complex $a + a_3$ has now been solubilised from mitochondria by the use of bile salts, it has not been possible to separate two components. The term cytochrome oxidase, i.e. the enzyme which makes the final transfer of hydrogen to oxygen itself, may therefore apply to the a_3 component of the complex or to a single entity at present denoted $a + a_3$.

The reaction with carbon monoxide is of particular interest and importance. Ferrous iron forms a carbonyl with CO which differs from other metal, for example Cu, carbonyls by being unstable to light. Warburg argued that, if, as he thought, the respiratory enzyme was using iron, its activity should be blocked by CO poisoning in the dark and the block removed by strong illumination. By measuring the amount of light-reversal at different wave lengths he was able to plot an action spectrum corresponding with the absorption spectrum of the oxidase-CO compound. In 1929 he published the results of an experiment with yeast, which gave a haem-type spectrum (Figure 24), similar, for example, to that of carbonyl haemoglobin with α, β, and γ peaks at 590, 540 and 430 nm respectively. It was later shown that cytochrome a_3-CO also

TABLE 5

Properties of the principal respiratory cytochromes

Cyt.	E_0'	Absorption peaks			MW	Reaction with		
	volts	α	β	γ	Monomer	CO	HCN	O_2
a	+0·29	603	none	452	66 000	No	No	No
a_3		603	none	448		Yes	Yes	Yes
b	−0·04	563	530	432	28 000	No	No	Slight
b_3	−0·06	560	529	425		No	No	Yes
b_5	+0·02	557	527	423		No	No	Yes
b_7	−0·03	560	529	—		No	No	Yes
c	+0·26	550	521	417	12 000	No	No	No
c_1	+0·26	553	524	418	40 000	No	No	No

has bands at 590 and 430 nm. This, of course, is not identical with the spectrum of cytochrome a_3 itself which has peaks at 603 and 448 nm; a β peak is not observable.

By far the strongest absorption is at 430 nm. If the respiration of a cell is passing over cytochrome a_3 as terminal oxidase, it is to be expected that it will be inhibited in an atmosphere containing a high CO/O_2 ratio and that the inhibition will be reversed by intense blue light. This has been shown to occur with a very wide range of cells from animal and plant tissues as well as many eucaryotic unicells and bacteria.

Carbon monoxide combines with ferrous iron and competes with oxygen; to obtain any considerable inhibition high CO/O_2 ratios are needed and values of about 20 are generally used experimentally. HCN and other reagents such as azide and H_2S, combine with the ferric iron and are therefore non-competitive.

Since their affinity for Fe^{3+} is high and the amount of iron present in the enzyme is small, more or less complete inhibitions can be obtained with small amounts of inhibitor. The inhibitors can therefore be used at low concentrations, usually around 0·001 M, at which they do not react with other cell constituents.

Cytochrome $a + a_3$, appears to have 1 or 2 atoms of Cu firmly attached; but the part, if any, played in oxidation is obscure. The cytochromes in general are usually regarded as carrying out oxidation-reduction by transporting electrons through their iron.

66) $$Fe^{3+} + e \rightleftharpoons Fe^{2+}$$

It has to be admitted, however, that the properties of the iron bound into cytochromes differ from those of iron in free solution. For example, only in cytochrome a_3 does it form carbonyls or react with HCN. In cytochrome c it will not react with oxygen and in most b-type cytochromes reacts only very sluggishly. It is therefore possible that other parts of the molecule are also involved in the working of the redox chain.

Bacterial cells contain cytochromes similar to but not identical with those of animals and plants. The cytochrome $a + a_3$ of *Bacillus subtilis*, for example, does not readily oxidise mammalian cytochrome c. Conversely, purified cytochrome c from *Azotobacter vinelandii* is not oxidised by mammalian cytochrome oxidase, or indeed by the oxidase from some other bacteria. There appear to be additional autoxidisable cytochromes in bacteria which have been named a_1, a_2 etc. There are also additional members of the other two groups, e.g. b_1, b_4, c_3, c_4 and c_5. These occur in varying combinations in different bacterial species, offering numerous possible variations for an electron transport chain.

Ubiquinone (Coenzyme Q)

Ubiquinone is a substituted benzoquinone with a polyisopentenoid side chain

In the substance from animal tissues $n = 10$; but in that from yeast $n = 7$ and many bacteria produce Q–6 and Q–8. Ubiquinones appear to be universal; but are particularly abundant in heart muscle. Among plants *Arum* spadix is a rich source. Ubiquinones occur in many parts of the cell and the amounts in the mitochondria alone exceed the amounts of other members of the respiratory chain about tenfold. Oxidation-reduction occurs at the quinol-quinone position.

67) [structure of benzene-1,4-diol (hydroquinone, OH groups)] $+ 2\ Fe^{3+}$ ⇌ [structure of benzoquinone (O groups)] $+ 2\ Fe^{2+} + 2H^+$

Electrons could be passed to a cytochrome. In Figure 23 it is suggested that CoQ may be reduced by flavoprotein (NADH$_2$, or succinic, dehydrogenase) and pass electrons to cytochrome *b*. This part of the chain is at present the least firmly established. The reasons why CoQ is supposed to participate in the chain are as follows. When it is removed from mitochondria by acetone their ability to oxidise succinic acid is lost, but can be restored by small additions of Q–10. The ubiquinone in mitochondria is reduced by succinate and reoxidised by atmospheric oxygen. Its position in the chain is suggested by the fact that antimycin A blocks the oxidation of the internal or added quinol and that reduction of ubiquinone by mitochondria is blocked by amytal when NADH$_2$ is the reductant; though not when succinate is (cf., Figure 23).

Ubiquinone has a strong absorption peak at 270 nm in the ultraviolet which is quenched in the reduced ubiquinol (Figure 24). Cauliflower mitochondria supplied with succinate show a reduction in light absorption between 250 and 300 nm with a maximum at 275 nm, and the decrease has been attributed to reduction of ubiquinone.

Oxidative phosphorylation

If oxidative phosphorylation were linked solely with the dehydrogenation of the substrate molecule, as glycolytic phosphorylation is, then only one molecule of phosphate would be esterified for each 2H transferred to $\frac{1}{2}O_2$. The phosphorylation ratio, P/O, would then be 1: but it was found by Belitzer in the 1930s that minced muscle preparations oxidising acids of the citric acid cycle, esterified approximately 2 P_i into phosphocreatine for each atom of oxygen consumed. He therefore suggested that there must be a type of phosphorylation associated with the oxidation of NADH$_2$. Numerous later experiments with preparations from animal and plant tissues and their mitochondria have shown that maximal P/O ratios are 4 for α-ketoglutaric acid, 3 for isocitric and malic acids and 2 for succinic acid. Intact mitochondria do not normally take up NADH$_2$ from an outside solution; but after a brief treatment with a hypotonic solution they will do so. After such treatment rat liver mitochondria have been shown to oxidise $NADH_2$ with P/O ratios approaching 3. The connection between respiratory chain oxidation and phosphorylation is uncoupled by dilute dinitrophenol. The extra phosphorylation with α-ketoglutaric is at the acid substrate level (p. 61) and survives such treatment; there is one 'missing' with succinic acid which starts its oxidation at a lower level.

Figure 23 (p. 77) indicates the three steps of the respiratory chain which represent a free energy drop that might be adequate to provide for a phosphorylation of ADP → ATP, i.e. whose ΔG_0 is in excess of about −9 kcal. One of these is $NADH_2$ → flavoprotein ($NADH_2$ dehydrogenase) which is missing from the succinate oxidation. The other two, cytochrome b → cytochrome c and cytochrome $a + a_3$ → O_2 are common to both paths. Experiments with inhibitors confirm the prediction. Rat liver mitochondria poisoned with antimycin A oxidise pure cytochrome c with a P/O ratio = 0·8. With HCN present to block the reoxidation of cytochrome c, β-hydroxybutyrate was oxidised via $NADH_2$ by added cytochrome c–Fe^{3+} (oxidised cytochrome c) with a P/2e ratio approaching 2·0.

For glycolytic phosphorylations it is possible to write the equations that link oxidation and phosphorylation together. So far, in spite of much effort, no common chemical intermediate has been identified in respiratory chain phosphorylation comparable with those in the anaerobic oxidation of glyceraldehyde-3-phosphate (p. 35). So far, phosphorylation has only been demonstrated in mitochondrial particles that retain some degree of the membrane organisation. An alternative line of explanation of the oxidation-phosphorylation linkage at present under intensive examination is therefore physical rather than purely chemical. During the transport of 2H from the substrate to oxygen, the protons and electrons appear to become separated, at least at some, e.g. cytochrome, stages (cf., also Eqn 67). If, as appears likely, the membrane is impermeable to H^+ and OH^- ions a pH difference might therefore be set up across it. A three-stage process corresponding to the oxidation of an acid substrate (SH_2) through NAD might be represented as in Figure 26.

The chemiosmotic hypothesis suggests that the excess of H^+ on the outer side of the membrane might lead to the formation of ATP by means of an ATP-ase driven into reverse. Briefly, mitochondrial ATP-ase has been shown to be so orientated that the reaction

$$68) \qquad ATP + H_2O \xrightarrow{\quad ATP\text{-}ase \quad} ADP + H_2PO_4^- + H^+$$

leads to an accumulation of H^+ at the outer surface. If respiratory chain oxidations are simultaneously producing H^+ at the outer surface, the outside H^+ concentration will be enhanced and at the same time the internal H^+ concentration will be reduced by the accumulation of OH^- (Figure 26). In the equilibrium

69) $$ATP + H_2O + H^+_{inner} \rightleftharpoons ADP + P_i + H^+_{outer}$$

a high ratio $[H^+]_{outer}/[H^+]_{inner}$, i.e. a large pH difference, tends to drive the reaction in the direction of ATP synthesis.

Experiments with rat liver mitochondria returned to air after a period in nitrogen showed an absorption of oxygen and a rapid build up of H^+ in the medium. If β-hydroxybutyrate, which is oxidised via NAD, was fed, a H^+/O ratio just below 6 was obtained; with succinate the ratio was approximately 4. This accords with the expectation that oxidation of $NADH_2$ would give rise to three separations of $2H \rightarrow 2H^+ + 2e$ (Figure 26) and that succinate would give only two.

INSIDE

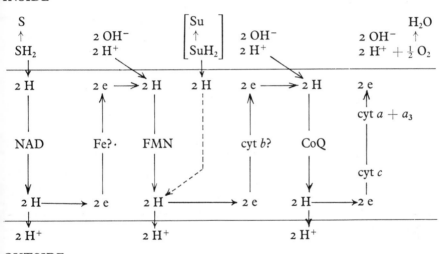

OUTSIDE

Figure 26   Diagram to illustrate the chemiosmotic hypothesis of charge separation across a mitochondrial membrane due to oxidation via NAD or oxidation of succinic acid. SH_2 substrate acid; SuH_2 succinic acid.

The converse experiment of causing the mitochondrion to synthesise ATP by lowering the external pH (increasing $[H^+]$) has not yet been convincingly achieved; but a corresponding experiment has succeeded with chloroplast lamellae. A pH difference, in this case with excess H^+ inside the lamellar sacs and excess OH^- on the outside, has resulted in the synthesis of ATP.

The efficiency of ATP formation

The respiration of one equivalent of glucose by way of glycolysis, the citric acid cycle and the respiratory transport chain appears to suggest a maximal yield of 36 ATP molecules per molecule of glucose. At $\Delta G_0 = -9$ kcal this represents a 'storage' of 324 kcal. The ΔG_0 for the complete oxidation of a mole of glucose is about -686 kcal, so that the efficiency of energy fixation in ATP is just under 48 per cent. This raises the question whether ATP is the only channel through which respiratory energy can be applied to useful purposes in the cell. So far as growth and the syntheses which it requires are concerned, no other is at present known. It has, however, been suggested that the energy causing the active uptake of salts into mitochondria, i.e. uptake against their concentration gradients, and some other cellular activities may be more directly dependent on electron transport, and that the total energy conserved is the sum of all these items. Little that is definite is yet known about them. The uncertainty about the means by which electron transport is geared to cell requirements, is probably the greatest gap in our present knowledge of respiratory mechanisms.

Terminal oxidases

The terminal oxidase of the so-called respiratory chain, i.e. cytochrome oxidase (cytochrome a_3, cytochrome $a + a_3$), is inhibited by cyanide at concentrations below 0·001 M; but the respiration of virtually all cells, as measured by oxygen uptake or CO_2-output, is resistant to extents varying from about 10–100 per cent. Occasionally, as in the fungal sheath of beech mycorrhizas, the respiration rate is even increased. The respiration of juvenile tissues may be strongly inhibited and show a progressive resistance as the tissue ages. The explanation of the discrepancy has proved difficult and complex. The most obvious likelihood, that cytochrome oxidase is not the only terminal oxidase, may apply in some cases.

The flavoproteins with attached metals, Mo, Mn, Cu or Fe, are all in varying degrees capable of reacting with oxygen. Their rate of reaction is much slower, however, and, whereas cytochrome oxidase is saturated at extremely low oxygen pressures, they require relatively high ones to give appreciable rates of reaction. The oxygen concentration for half-saturation of yeast cytochrome oxidase is probably below 10^{-7} M, whereas the corresponding figure for the flavoprotein oxidases is around 10^{-3} M.

The respiration of the muscle cells of *Ascaris lumbricoides*, an intestinal parasite, is not inhibited by cyanide, azide or antimycin A. Particulate preparations are able to oxidase succinic acid and $NADH_2$, but do not contain any cytochrome c or cytochrome oxidase. They require high oxygen concentrations for full activity and are activated by Mn^{2+}. The terminal oxidase has been extracted and when reduced with succinic acid showed the difference (oxidised–reduced) spectrum to be expected of a flavoprotein. The main constituent

vas shown to be FAD. It is, however, doubtful how far the parasite depends
on this system for its respiration, since it normally exists anaerobically (p.
51).

Probably the most carefully examined case of high resistance to HCN inhibition is that of the aroid spadix (Figure 27). This is notable for its very fast

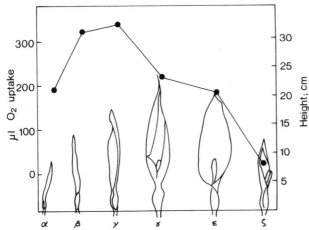

Figure 27 Stages in the development of an *Arum* inflorescence. Height in cm on the
right. Oxidative respiration of 15 spadix slices shown by the curve and the scale at
left. The club-shaped spadix is fully exposed at ϵ and withered at ζ.

respiration rate which is not fully evident from Q_{O_2} values, because the cells
at their most active stages are tightly packed with starch. This is wholly consumed and there is a massive evolution of heat which leads to a rise of temperature within the enclosing spathe.

Investigation has revealed three enzymes or carriers capable of reacting
directly with oxygen; a flavoprotein, cytochrome $a + a_3$ and a cytochrome b_7.
The flavoprotein, which has FAD as its prosthetic group, appears to be ruled
out as a major direct path to oxygen, by the fact that maximal rates of respiration can be achieved at low oxygen pressures. It has been shown to function as
an NADH$_2$ dehydrogenase, linking with the cytochromes, perhaps through
ubiquinone which is also present (cf., Figure 28).

It has been shown spectroscopically that, in the presence of cyanide, cytochrome a_3 remains reduced whereas the b cytochromes and cytochrome c are

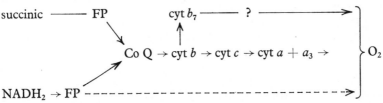

Figure 28 Oxidation chains in the respiration of *Arum* spadix.

oxidised. Since cytochrome a_3 reacts with CO, cyanide and similar inhibitors and cytochrome b_7 does not, it might appear a clear choice in favour of the b_7. The possibility remains, however, that the b_7 is only acting as a shunt to oxygen when the usual route is closed; there are also other possibilities.

Extracted mitochondria fed with α-ketoglutarate have been shown to form ATP with a high P/O ratio which is reduced by cyanide or azide poisoning. This might be expected when the cyt $b \to$ cyt $c \to$ cyt $a + a_3$ steps are cut out; but is no guarantee that they were functioning in the intact cell. Nothing is known of possible intermediates in the oxidation of cytochrome b_7 or their conceivable connection with phosphorylation. All the systems mentioned are present in the mitochondria, and their possible interrelations are indicated in Figure 28. At present it remains a matter of speculation to what extent and in what ways the aroid respiration chain differs from the conventional cytochrome series. In its earlier stages the respiration appears to go through normal glycolysis followed by a citric acid cycle. Its special feature is the possession of an unusually high content of b-type cytochromes, particularly cytochrome b_7, introducing the possibility of a cytochrome shunt which may be less fully associated with phosphorylation. This might be linked with the fact that the aroid spadix is a sterile tissue running down to exhaustion with a rapid heat production.

It is not known to what extent similar shunts occur in other cells. They may be associated more with plants, in which cytochrome b_3 is also known, than with animals. At present perhaps the most important effect of the aroid results is to emphasise once more the complexity of cellular systems and the difficulty of deciding what they usually do as distinct from what they are capable of doing.

Further Reading

Beevers, H. (1961) *Respiratory metabolism in plants*, Chapter 4. Row, Peterson, New York.

Bendall, D. S. and Bonner, W. D., Jnr. (1971) Cyanide-insensitive respiration in plant mitochondria. *Plant Physiol.*, **47**, 236–45.

Lehninger, A. L. (1964) *The Mitochondrion*, Chapters 4 to 6. Benjamin, New York.

Lundegårdh, H. (1960) The cytochrome-cytochrome oxidase system. *Encyclopaedia of Plant Physiology*, **12**, 311–64.

Morton, R. A. (1965) Quinones as biological catalysts. *Endeavour*, **24**, 81–6.

Robertson, R. N. (1968) *Protons, electrons, phosphorylation and active transport.* University Press, Cambridge.

Slater, E. C. (1960) Oxidation-reduction potentials and their significance in hydrogen transfer. *Encyclopaedia of Plant Physiology*, **12**, 114–31.

Smith, L. (1961) Cytochrome systems in aerobic electron transport. In *The Bacteria*, 2nd edn. vol. **2**, 365–96. Academic Press, New York.

6 Special Oxidations

THERE are a considerable number of enzymes in cells that react directly with molecular oxygen. Some of them have been regarded as terminal oxidases of respiration transport chains replacing or accompanying the cytochrome chain. Whether they behave in this way or not, there is also the question of more direct oxidations, not directly linked with the main stream of respiration. For example, the hydroxylation of a substituted benzene to the corresponding phenol

70) $\qquad \xrightarrow{+\frac{1}{2}O_2} \qquad$

is a commonly occurring reaction which requires oxygen and may proceed directly through a specific oxidase.

The phenol oxidases

The phenol oxidases are probably a large group of enzymes; but owing to the multiplicity of their sources, from the melanocytes of human skin to the mycelia of fungi, and to the well-nigh unlimited number of their substrates, specificity has proved difficult to determine. They seem however to fall into two groups, the cresolases and the laccases.

The cresolases have a twofold reaction, which with p-cresol as substrate may be written

71) $\qquad \xrightarrow[\text{cresolase}]{+\frac{1}{2}O_2} \qquad \xrightarrow[\text{polyphenolase}]{+\frac{1}{2}O_2} \qquad + H_2O$

The cresolase and polyphenolase activities may be due to two separate enzymes, but so far they have defied separation, and are usually regarded a depending on two sites on the same enzyme. The cresolase activity is slower than the polyphenolase so that starting from a monophenol there is a lag phase which can be abolished by adding a catalytic amount of diphenol. Both activities require the presence of firmly bound copper and may be inhibited by cyanide and by CO. The copper carbonyl is not, however, dissociated by light and the CO inhibition is not light reversible. Inhibition by cyanide is often less marked than the inhibition of cytochrome oxidase. Conversely, the copper-chelating compound diethyldithiocarbamate (DIECA) strongly inhibits phenol oxidases and hardly affects cytochrome oxidase, and may, with adequate precautions, be used to discriminate between their activities.

Laccases have no monophenolase activity and although they also contain copper are not inhibited by carbon monoxide. They are typified by the laccase of *Rhus vernicifera* which oxidases urushiol,

$$
\begin{array}{c}
\text{OH} \\
\bigcirc\text{OH} \\
\text{C}_{15}\text{H}_{27}
\end{array}
$$

The quinones formed as the primary products of phenolase activities are highly active both in further oxidation and condensation with, for example, amines and amino acids.

With catechol as substrate, the oxygen consumption *in vitro* is equivalent to 1 molecule O_2 per molecule catechol, due to a further hydroxylation

72)

$$
\begin{array}{ccccc}
\text{catechol} & + \tfrac{1}{2}O_2 & \longrightarrow & \text{o-quinone} & + \tfrac{1}{2}O_2 & \longrightarrow & \text{p-hydroxy-o-quinone}
\end{array}
$$

and the p-hydroxy-o-quinone so formed may condense with amines to form corresponding p-amino-o-quinones. Continuous oxidation of the amino acid glycine to glyoxylic acid $+ NH_3$ may be set up by this route. The initial amino-quinone need not be derived from glycine, and is coloured variously according to the amine used. An immense variety of compounds may be derived, and CO_2 may or may not be released according to the phenols and amines involved. One of the most interesting examples is the oxidation of tyrosine, which contains both phenolic and amino groups in its own molecule. It was first extensively studied by Raper with the tyrosinase of mealworm. The initial reaction is a cresolase type oxidation of the tyrosine to dihydroxyphenylalanine (dopa)

73)

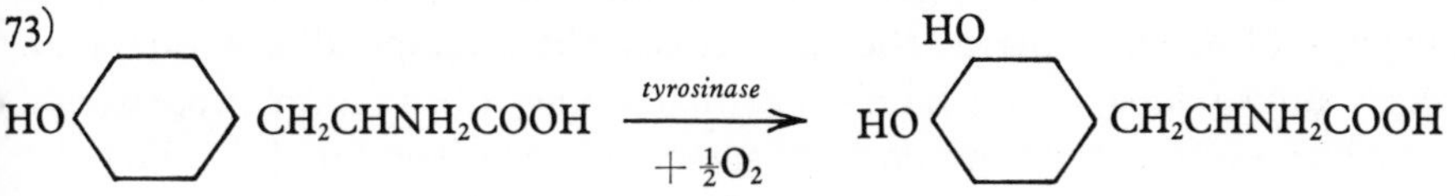

$$HO\langle\bigcirc\rangle CH_2CHNH_2COOH \xrightarrow[+\frac{1}{2}O_2]{tyrosinase} HO\langle\bigcirc\rangle CH_2CHNH_2COOH$$

dopa

followed by a polyphenolase oxidation to dopaquinone, and rapid, possibly non-enzymatic, ring closure, and decarboxylation leading to a yellow coloured indole quinone.

74)

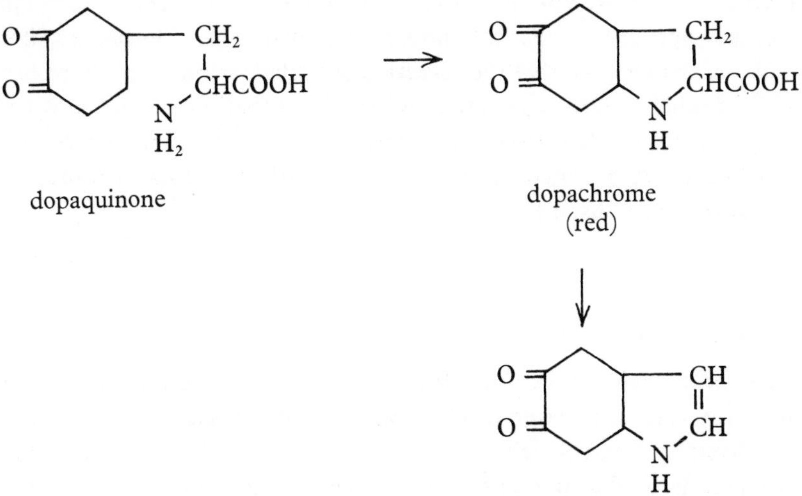

dopaquinone

dopachrome
(red)

indole-5,6-quinone
(yellow)

Polymerisation of the indole quinone, probably associated with other reactions, leads to the formation of melanins which are the black pigments formed in melanocytes. The pigments accumulate in rods or spheres with diameters around 0·3 μm. These granules have a high protein content and contain a variety of enzymes. The melanins appear to be bound to proteins through sulphur linkages.

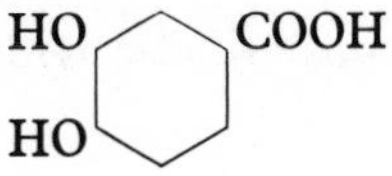

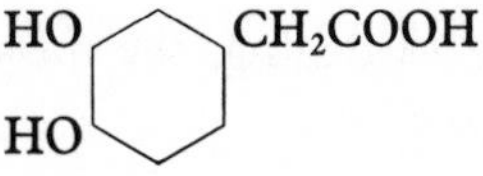

All the black, brown and buff pigments formed in the skins of higher animals are of this nature but, among plants, only the black spots on broad bean flowers are known to have a similar origin. The darkening of insect cuticles is also due to melanin formation and is accompanied by a second process also due to phenolase activity. Polyphenols such as dihydroxybenzoic acid and dihydroxyphenylacetic acid, occur in the outer cuticle together with soluble and lipo-proteins. Under the influence of polyphenolases, they give rise to the corresponding o-quinones, which, unlike dopa, have a tanning, i.e. hardening, effect on the proteins and so give rise to the hard and dark substances of the outer cuticle. Similar reactions also occur in hair and feathers; but in all these latter the reactions probably go on outside the living protoplasm.

Bacteria, for example *E. coli* and *Streptomyces scabies* that causes potato scab, secrete phenolases into the external medium. The same is done by the living mycelia of many higher fungi and by their overripe fruit bodies (mushrooms, etc.). What role they play in the life of the cells is obscure; but in the soil they appear to assist in humus formation and so to improve soil fertility. It has been proposed that the excretion of such enzymes may afford protection against other micro-organisms; and similarly that the phenolase activity that occurs in many plant tissues damaged by fungal invasion affords protection against further advance by the fungi. In neither case has a protective effect been firmly established.

Lacquer formation

The most obviously striking feature of phenolase activity is the extent to which it runs wild after mechanical or other damage to living cells; or at senescence, when the cell structure is beginning to break down. This is particularly noticeable among the higher plants, where it has at least three consequences of commercial importance. In the first place it may lead to undesired discolorations in food products, such as dried fruits, and must be prevented by bleaching for example with sodium bisulphite. Conversely, it may be the basis of a process such as the production of Japanese lacquer. The latex of *Rhus vernicifera* and, in China *R. silvestris*, is obtained by tapping the bark and scraping out the small yield of fluid. The raw latex is a thick greyish emulsion which is stored in closed vessels. It is filtered through cotton wool or hemp and applied in successive very thin layers. Oils, gold, silver and other pigments may be added before application. The hardening and blackening occurs *in situ*, due to the laccase acting on diphenols, the best known of which are urushiol and laccol.

During the hardening, the unsaturated side-chains are also hydrogenated as in a drying oil.

Tea fermentation

Tea fermentation is also very largely an unorganised phenolase reaction. The plucked leaves, unopened tips and first two or three leaves, are first withered by being left on trays in the open for a day, and are then passed repeatedly through rollers. The pressure does not break up the cell walls, but disorganises the protoplasm. The leaves kept at a temperature of 20°C–27°C in a humid atmosphere become reddish brown and slowly darken as the phenolases act on tea tannins. After 3 or 4 hours the enzyme activity is stopped by 'firing', i.e. by drying at about 65°C in a heated chamber until the moisture content is reduced to 3 or 4 per cent. In green teas the leaves are heated before browning has occurred; but are afterwards allowed to ferment. They retain their green colour as most of the phenolase has been inactivated by the initial heating. The caffeine (trimethylxanthine) to which tea owes its stimulant action, occurs largely in combination with tannins and is at least partly released during fermentation. It has been conclusively shown that tea fermentation is independent of micro-organisms and can be reproduced with the extracted leaf enzymes. The leaf tannins are a very complex mixture that has been resolved by chromatography. Epicatechin and its gallate are typical examples

The tea enzyme appears to have no cresolase activity, but is highly active, especially with its native tannins, as polyphenolase. The o-quinones formed condense giving rise to still more complex tannins.

The attempt to assess the normal behaviour of phenolases in living cells is made the more difficult by their wide intracellular distribution. They are frequently readily soluble and probably located throughout the soluble cytoplasm; but there is also frequently a fraction which is virtually unextractable. In tea leaves this appears to be located in the chloroplasts. In dermal melanocytes the activity leading to melanin formation is located in organelles resembling mitochondria.

Cellular function of phenol oxidases

The o-quinones which are the product of polyphenolase reaction have high positive redox potentials and will readily oxidise such natural reductants as ascorbic acid and cytochrome *c*. Smooth-running redox systems may be reconstructed *in vitro* which will oxidase $NADPH_2$; no complicating condensations occur at the quinonoid level until the reductant is exhausted. It has often been suggested, particularly in connection with higher plant tissues, that such systems form respiratory transport chains, analogous to or supplementing the cytochrome system. In spite of a great and diversified amount of work no unequivocal answer has yet been attained.

Aerobic hydroxylation of aromatic compounds is widespread in cells of all types. Cresolase activity is the oxygen-consuming hydroxylation of a monophenol to the corresponding o-phenol (p. 91, Eqn 71). For continuous operation a reductant is required which, in a phenolase complex, may be a diphenol or other reducer capable of keeping the enzymic copper in the cuprous state.

The conversion of tyrosine to dopa (p. 93, Eqn 73) is an example. In plant cells a vast range of o-diphenols exists among the tannins, flavonoid and other phenolic compounds and their formation may be an important cell function of the phenolases. In liver and other animal cells there are aerobic hydroxylases which form monophenols. One of the best known is phenylalanine p-hydroxylase, found only in liver, which forms tyrosine from phenylalanine. This enzyme also consumes oxygen and requires a soluble cofactor, a pteridine,

which must be kept reduced by $NADH_2$ or $NADPH_2$. The reduction requires a pteridine reductase. A genetic block prevents the formation of phenylalanine hydroxylase in some humans. The resulting metabolic abnormality, phenylketonuria, does not lead to much physical disability, but causes gross mental retardation.

The aerobic hydroxylases afford means by which oxygen is used for specific metabolic oxidations, which may be synthetic, as in tannin formation, or degradative as in phenylalanine disposal. They illustrate the fact that transfer of hydrogen (or electrons) to air is not limited to a single respiratory channel.

Ascorbic oxidase

Ascorbic acid appears to be formed universally in higher plant cells and in many fungi and bacteria. It is a required vitamin C, in animal cells. It can be oxidised by Cu^{2+}, free or associated with proteins, the cytochrome and phenolase systems, and by a specific Cu-protein, ascorbic oxidase, of relatively limited distribution.

75)

$$
\begin{array}{ccc}
& O & \\
& \parallel & \\
& C & \\
HO-C \diagup & \diagdown & \\
\parallel & & O \\
HO-C \diagdown & \diagup & \\
& C & \\
& \diagup \diagdown & \\
H & CHOH & \\
& | & \\
& CH_2OH &
\end{array}
\quad
\begin{array}{c}
-2H \\
\rightleftharpoons \\
+2H
\end{array}
\quad
\begin{array}{c}
O \\
\parallel \\
C \\
O=C \diagup \diagdown \\
| \quad\quad O \\
O=C \diagdown \diagup \\
C \\
\diagup \diagdown \\
H \quad CHOH \\
| \\
CH_2OH
\end{array}
$$

ascorbic acid   dehydroascorbic acid

A dehydroascorbic reductase is also known from plants which links the system with glutathione and, through it, to $NADPH_2$. A second system, independent of glutathione, but probably also involving a flavoprotein as reductase, oxidases $NADH_2$ (Figure 29).

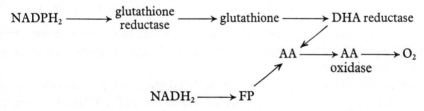

Figure 29   Oxidation chains involving ascorbic acid.

Such series have been assembled *in vitro* and have been supposed to play an occasional rôle in respiratory oxidation. Ascorbic oxidase is little affected by CO poisoning; but is strongly inhibited by DIECA. It has been noticed that, during an early stage in the germination of some barley seedlings, the cytochromes in their root cells remain permanently oxidised. During the same period their respiration ceases to be inhibited by CO and becomes sensitive to DIECA. Parallel experiments showed no corresponding stage in wheat seedlings. Many plants, when they are infected by parasitic fungi, show an enhanced oxygen uptake which is accompanied by increased ascorbic oxidase activity, and it has been suggested that the extra respiration passes at least in part over an ascorbic oxidase system.

Ascorbic oxidase is usually a readily soluble enzyme; but is sometimes very firmly attached to the cell wall and has also been reported in mitochondria. Its solubility and apparent ability to link with $NADPH_2$ oxidation suggests a possible connection with the pentose phosphate pathway (p. 71); but it has proved as difficult to assess the respiratory significance of ascorbic oxidase as that of the phenolases.

The very wide occurrence of ascorbic acid itself does not necessarily indicate a rôle as member of an oxidation chain for it. Under normal circumstances the ascorbic oxidase in cells is almost wholly in the reduced form and it may provide a buffer of reducing power rather as glutathione is said to do in red cells (p. 71).

Glycollic oxidase

A direct oxidase for glycollic acid has been found in a wide variety of cells. It has been isolated and purified from liver and from spinach leaves. It is a flavoprotein with FMN as its prosthetic group, and is mainly attached to organelles, called peroxisomes or glyoxysomes (Plate IX) of much the same size range as mitochondria but slightly denser and without the characteristic internal structure of mitochondria. Hydrogen peroxide is a product of the reaction

$$76) \quad CH_2OHCOOH + O_2 \xrightarrow{\text{glycollic oxidase}} CHOCOOH + H_2O_2$$

and in the presence of catalase, which is also located in the peroxisomes, the hydrogen peroxide is decomposed to water and oxygen or could conceivably oxidise other substances.

A glyoxylic reductase has also been isolated from some tissues which reduces the glyoxylic acid back to glycollic at the expense of $NADH_2$. It has been located in the peroxisomes of spinach leaves.

Photorespiration

Experiments involving the use of $^{14}CO_2$ have led to the conclusion that the photosynthetic cells of some leaves and algae give off CO_2 a good deal faster in the light than in the dark. The effect increases with oxygen concentration and may continue right up to 100 per cent oxygen in the atmosphere; it is suppressed anaerobically. The response to oxygen differentiates the extra CO_2-output very sharply from the normal respiratory CO_2 production of plant cells. There is reason to suppose that it results from the metabolism of glycollic acid.

Glycollic acid appears to be a normal product of photosynthesis, especially in low concentrations of CO_2, such as are usual in the atmosphere. The administration of α-hydroxysulphonates, which inhibit glycollic oxidase, leads to a marked accumulation of the acid. The oxidase, which as a flavoprotein has a low oxygen affinity, is particularly abundant in the green cells of leaves and some of it is always in the chloroplasts. Isolated chloroplasts of the alga *Acetabularia* are apparently able to carry out the entire sequence of photorespiration, producing extra CO_2 in the light. It has been noted, however, in preparations from spinach leaves that addition of peroxisomes to the extracted chloroplasts greatly accelerated the production of CO_2 from glycollic acid, though the glycollic oxidase of peroxisomes oxidises glycollic acid only to glyoxylic acid (Eqn 76 above) and not to CO_2. Not all green cells show photorespiration; it is notably absent from maize and sugar cane leaves whose CO_2 fixation paths under normal conditions are different from the commonly quoted one and which may not accumulate glycollic acid. Glycollic acid may be utilised in many cells in the synthesis of sugars and amino acids.

Further Reading

Goldsworthy, A. (1970) Photorespiration. *Bot. Rev.*, **36**, 321–40.
Kisaki, T. and Tolbert, N. E. (1969) Glycolate and glyoxalate metabolism by isolated peroxisomes or chloroplasts. *Plant Physiol.*, **44**, 242–50.
Malkin, R. and Malmström, BoG. (1970) The state and function of copper in biological systems. *Adv. in Enzymol.*, **33**, 177–244.
Mason, H. S. (1955) Comparative biochemistry of the phenolase complex. *Adv. in Enzymol.*, **16**, 105–84.
Pridham, J. B. (Editor) (1963) *Enzyme Chemistry of Phenolic Compounds.* Pergamon Press, Oxford.

7 Oxygen Supply

The invasion coefficient

Most cells consist of 75 per cent or more of water, which is liable to evaporation. They must therefore live in an environment of high humidity. Most free-living unicells live either in water or in the body fluids of a host. Respired oxygen is always dissolved oxygen, because even when it is brought to the cell in a moist gas phase it must be dissolved in the water of the cell surface before reaching the oxidases. This circumstance is significant because oxygen has a low solubility, and a low rate of diffusion in water. Over the biological range of temperatures solubility decreases at about the same rate as diffusion increases; with the result that the invasion coefficient of oxygen into a cell changes very little with temperature. The rate of respiration, i.e. the oxygen consumption rate, however, approximately doubles for 10°C rise so that, particularly in the upper permissible ranges, it might be supposed that diffusion from the cell surface to the respiratory oxidation centres could be a rate-determining factor.

The oxidases of bacteria are almost certainly located at the cell surface (p. 8); in highly vacuolated cells, the mitochondria are of necessity pressed against the cell wall; but, in general, there is no clear indication that the respiratory oxidases are so arranged in the cell structure as to keep diffusion paths to a minimum.

Yeast cells have an outstandingly high respiration rate and an approximately spherical form. The drop of oxygen concentration to be expected at the centre may be calculated using data as in Tables 1 and 3 and Gerard's equation for the invasion of a sphere

$$77) \qquad U = C - \frac{a(R^2 - r^2)}{6D}$$

where C is the external concentration of oxygen in atm, R the radius in cm, D the invasion coefficient (ml at NTP/cm² min for gradient 1 atm/cm). At the centre, $r = 0$; D may be taken $= 3.4 \times 10^{-5}$, the value for water, and the upper limit of respiration in Table 3 gives $a = 0.042$ in the appropriate units. The second expression on the right hand side of the equation then $= 3 \times 10^{-5}$ i.e. the external concentration of oxygen, 0.21 atm, is reduced by only a negligible amount at the centre of the cell.

The rate/[O$_2$] hyperbola

Nevertheless, oxygen supply may exercise a restriction on rates of aerobic respiration. Apart from flavoproteins, the affinities of the oxidases for O_2 are high. The K_m (O_2-concentration for half-saturation) of yeast oxidases is said to be of the order 10^{-6} atm so that maximal oxidation rates may be achieved at very low concentrations of oxygen, provided that the concentrations are continuously maintained at the cell surface. This last condition is not, however, met when a unicell is respiring in still water and, owing to the slow rate of diffusion of oxygen in water, diffusion shells are set up with vanishing concentrations of oxygen at the cell surface. A cell embedded in a massive tissue is at a still greater disadvantage, because the surrounding cells are themselves using oxygen. Attempts to determine the relation between O_2-uptake and external concentration, with a wide variety of cells from bacteria to animal and plant tissue cells, have yielded a variety of hyperbolas with half maximal rates at very varying positions. Although such curves may resemble the rate/substrate concentration relationship of many enzyme reactions, it is unlikely that any of them are determined solely by the reaction of the terminal oxidase with oxygen. Even with relatively high external diffusion resistances, it does not follow that oxygen supply will become limiting at normal atmospheric pressure. In pieces of excised tissue (liver, etc.) this may happen, as the blood supply has been cut off. Similarly in pieces of plant tissue at relatively high temperatures, respiration rates may tend to outrun oxygen diffusion through the intercellular spaces. The coefficient of diffusion of oxygen through potato tissue at 25°C has been calculated at about 2.9×10^{-4} ml/cm^2 sec, and the respiration rate as 1.6×10^{-6} ml/cm^3 sec. With these parameters, to reduce the central oxygen concentration from atmospheric to 2 per cent would require a very large potato indeed—about 30 cm across. But the situation changes rapidly when the temperature is pushed up another ten degrees. The reason is that while the rate of respiration is about doubled, the rate of invasion by oxygen may not rise at all (cf., p. 100). Potatoes kept at around 35°C soon develop 'black heart' involving complete break down of the central tissues.

Cells embedded in animal tissues are saved from a similar fate by the fact that oxygen is carried to them by mass movement in the highly specialised red cells of the blood stream. These cells, so highly specialised to serve the respiratory requirements of others, have themselves an unusual respiration and consume virtually none of the oxygen they are transporting.

The red cell

Erythrocytes are formed in red bone marrow. They pass through a number of developmental stages containing increasing quantities of haemoglobin. At a relatively late stage of their development the nucleus begins to shrink. Thereafter no further division of that particular cell can occur, and when it enters the stream it no longer has either nucleus or mitochondria. Its surface consists of a membrane; if this is ruptured, the haemoglobin escapes leaving a 'ghost' consisting of about one fifth lipids and four fifths protein. The lipids are probably limited to the surface membrane; but the protein may also form a fine internal meshwork.

The rate of oxygen consumption by a red cell is exceedingly small; it has been estimated at about one thousandth that of a leucocyte. The red cells can glycolyse slowly and form a small amount of ATP by this means. This ATP is involved in the active transport of K^+ and other ions, and is one of the few examples where the physiological role of anaerobically produced ATP is definitely known. It also saves the mature erythrocyte from the reproach of being 'a mere haemoglobin-freighted drop of protein'. As already mentioned the red cells also contain enzymes of the pentose phosphate pathway, though they probably do not lead to oxidative ATP formation, but rather to maintenance of the membrane (p. 71).

Haemoglobin is an iron-porphyrin-protein akin to the cytochromes. The haem pigment has the configuration shown in Figure 25 (p. 82). The entire haemoglobin molecule has a molecular weight of about 67 000 and, thanks to the work of Perutz and his associates, its protein moiety is one of the very few whose amino acid sequences and three-dimensional structure are known in considerable detail. It consists of four spiralled polypeptide chains, each of which has a haem group recessed within its windings and each of which is similar to the chain of myoglobin (Figure 30) though their composition varies. Each haemoglobin molecule contains two pairs of identical polypeptides. There are four possible chains in human haemoglobins, which in varying combinations give rise to the different blood groups. Owing to the coiling of the peptide chains and their tetrahedral arrangement one to another the molecule as a whole is 'globular'.

The haem groups carry oxygen attached to their iron. In the presence of relatively high oxygen concentrations, as in lung capillaries, oxygen displace a water molecule from the iron; and, where oxygen concentration is low, the oxyhaemoglobin dissociates off the oxygen which is again replaced by a water

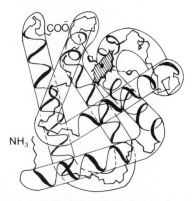

Figure 30 Diagram of the polypeptide chain in the myoglobin molecule, showing the straight helical regions and the non-helical foldings. The haem group is shaded. (After Kendrew, Dickerson, *et al*, simplified.)

molecule. Haemoglobin is half saturated at O_2-pressures around 0·03 atm, and is thus capable of operating at very low oxygen tensions. These oxygenation changes do not involve a change in the valency of the iron which remains in the ferrous state throughout. The dissociated oxygen passes by aqueous diffusion from the red cell through the capillary walls to the oxygen consuming cells of the receiving tissue.

Erythrocytes are small as cells go (Table 1, p. 5). This was known to Leeuwenhoek in the seventeenth century, and he used the size of the human blood corpuscle—about 1/3000 of an inch—as a standard for microscopic comparisons. Their high ratio of surface to bulk facilitates their oxygen exchanges; but in fact they often pack into rouleaux. Attention has been drawn to the fact that human erythrocytes take on the shape of a dimpled disc (Figure 31) in serum; a form which tends to provide maximal surface with adequate stability. In other vertebrates the shape is nearer to a simple disc. The shape is dependent on the external medium and, on addition of water, for example, the cells

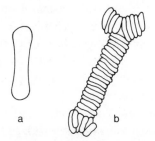

a   b

Figure 31 (*a*) Transverse section of red cell in serum; (*b*) a rouleaux. Diameter of each cell about 8μm.

become spherical without change of volume, but with corresponding contraction of surface area.

The respiratory function to which red cells are so highly adapted is thus to receive and donate dissolved oxygen at points of high and low oxygen pressure respectively, and so to provide occluded cells with free access to oxygen. They are carried passively in the blood stream from one point to the other, and themselves have a well-nigh negligible respiration.

Yeasts

Most cells can survive for a time, sometimes fairly prolonged, without access to oxygen (cf., p. 50). Very few, however, can grow and multiply without it. The classic example is yeast, or, more correctly, certain strains that can grow both aerobically and anaerobically. It cannot be over-emphasised, however, that even these yeasts grow more slowly and multiply less freely without oxygen than with it. The ceiling concentration of yeast cells in an anaerobic culture is much lower than in an aerobic one otherwise identical. The amount of sugar disappearing from the culture may be the same in both, but a much larger proportion of it, perhaps as much as three quarters, is incorporated into living yeast in the aerobic. This was noted by Pasteur. As he put it: in air, yeast behaved just like any other fungus. He also noted that with varying concentrations of oxygen there was a progressive change from the purely anaerobic fermentation with minimal growth in nitrogen to the more usual type of aerobic respiration and rapid growth in air. Since Pasteur's time it has been found that this change can be put into reverse by increasing the concentration of glucose in the culture. Up to a concentration of about 6×10^{-3} M glucose, the aerobic respiration of a young culture tends to increase with glucose concentration; but above that it is suppressed in favour of fermentation. This effect is peculiar to glucose and yeasts will respire substances such as lactic or acetic acids (which cannot be 'fermented') and even other sugars such as fructose, mannose, galactose or melibiose, which are fermented progressively more slowly in that order, without suppression at higher concentrations. It cannot yet be said that the biochemical mechanisms of either the suppression of fermentation by respiration (Pasteur Effect) or of the glucose suppression of respiration by fermentation (Crabtree Effect) is yet properly understood. The initial step, whatever it may be, leads to complex series of adjustments both of metabolism and structure.

Strains of *Saccharomyces cerevisiae* growing and respiring rapidly in air develop spherical or ellipsoidal mitochondria with plate-shaped cristae. This happens when they are growing on moderate concentrations of glucose, or on non-fermentable substrates, such as acetate or melibiose (Figure 32a). The mitochondria bear the usual cytochromes and other respiratory enzymes; and appear to operate a 'normal' electron transport chain with accompanying phosphorylation. Cells grown anaerobically and those grown in air, but with glucose repression, develop few mitochondria and those with few, if any, cristae (Figure 32*b* and *c*; Plate XIII). Cytochromes are not altogether lacking, but are present in much reduced amounts. There is said to be a close correlation between glucose repressibility and instability of the mitochondria. Aerobic yeasts, which do not grow anaerobically, do not readily change their normal mitochondrial form. Repression is reversible i.e. glucose-repressed yeasts on being transferred to, say, melibiose develop mitochondria carrying a normal cytochrome complement.

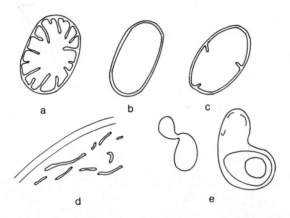

Figure 32 Forms of yeast mitochondria; (*a*) aerobic, on melibiose etc; (*b*) and (*c*) aerobic, glucose repressed, or anaerobic; (*d*) anaerobic, unenclosed membranes; (*e*) cytoplasmic mutant 'petites'. All except (*d*) single mitochondria. Diagrammatic.

Correlated biochemical and electronmicroscopical studies of the yeast *Torula utilis* have shown similarly that mitochondria are developed only in the presence of oxygen. Anaerobically grown cells on the other hand, while having no mitochondria, showed an unenclosed lamella-system consisting of single unit membranes lying more or less parallel with one another. A fraction was isolated centrifugally that presumably derived from these membranes. It had no powers of aerobic oxidation and contained no cytochrome $a + a_3$, but did contain very small amounts of cytochrome *b* and *c* and dehydrogenases for $NADH_2$ and succinic acid. Such anaerobic cultures supplied with air developed the ability to respire, and at the same time the unenclosed membranes fused apparently to form rudimentary mitochondria with a few cristae and there was a considerable development of cytochromes *b* and *c*. Anaerobic *Saccharomyces cerevisiae* also develops a system of parallel membranes below the plasmalemma which may have a similar function (Figure 32*d*).

The well-known 'petite' mutants of *S. cerevisiae* are unable to grow in air upon non-fermentable substrates. On fermentable sugars they grow slowly, producing small colonies on solid media. In the presence of 3 μg/ml acriflavine, almost 100 per cent of the daughter cells budded off by a wild type yeast may be petite type mutants; the mutation appears to be irreversible. Although unable to respire aerobically 'petites' produced by chromosomal mutation produce more or less normal looking mitochondria; those produced by cytoplasmic mutation or both types together produce mitochondria of a kind, often with whorled membranes (Figure 32*e*). They contain cytochrome *c*; but cytochromes *a* and *b* are both lacking and also dehydrogenases for $NADH_2$, lactic and succinic acids. Although oxygen is accessible, a respiratory block is developed nevertheless. Similar blocks are also known in the fully aerobic fungus, *Neurospora crassa*. The mutant 'poky' actually produces cytochrome *c* in amounts in excess of the normal; but is deficient in cytochromes *a* and *b*.

The limited growth achieved by the various types of respiratorily restricted yeasts is generally supposed to depend on the ATP formation in their anaerobic oxidation of glyceraldehyde-3-phosphate and the conspicuous development of cytoplasmic lamellae already mentioned may provide a special locus for this. With most cells it is difficult to show in what way this particular ATP is utilised.

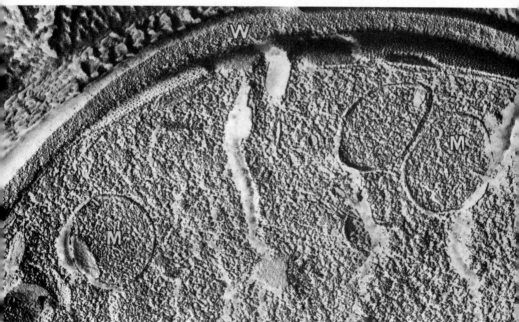

Plate XIII Mitochondria (M) in freeze-etch preparation of part of an anaerobic yeast cell. Mitochondria have walls but no cristae; W wall. $\times$ 44 000. Courtesy H. Moor.

Gas exchange and food storage

Meat (muscle cells) can be frozen and preserved for long periods in cold storage; seeds, garden peas for example, also have a low water content, 10–15 per cent, and can also be kept in the deep freeze. Owing to the high water content in its cell vacuoles, (around 85 per cent), most fruit is not frozen for storage because on rethawing it breaks down to a mushy pulp. It therefore has to be stored live and passes through successive stages of maturation, which may be autocatalysed, e.g. by the production of traces of ethylene ($CH_2{=}CH_2$). The problem resolves itself into slowing these stages down as far as possible without radically altering them. A given variety of fruit in passing from the green to the ripe condition emits a more or less fixed amount of carbon dioxide, i.e. its cells perform a given amount of respiration. Anything which tends to reduce the cellular respiration rate, tends to prolong the storage life, i.e. the transition from mature green to eating-ripe. Within limits, a reduction of temperature will do this; but every fruit has a critical temperature below which it will not ripen, however long it is kept. For bananas this is as high as 12–13°C. Temperate fruits tend to have lower limits; but even for a given sort, the minimal temperature may vary with variety. Bramley's seedling apples will not ripen below 4°C whereas Worcester Pearmains will ripen at 1°C.

The rate of respiration of most fruits is reduced by lowering the oxygen concentration around them; but again there is a lower limit below which ripening may not occur at all. At oxygen concentrations below about 5 per cent, anaerobic respiration is not completely suppressed. There is often a critical oxygen concentration below which CO_2-emission begins to rise again (Figure 33); though some fruits, for example avocados, do not show this. For

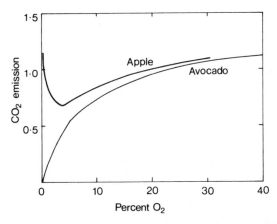

Figure 33 CO_2-emission of apples and avocados in varying concentrations of oxygen Rates in air = 1·0.

most fruits studied the optimal oxygen concentration for prolonging the storage life is around 2·5 per cent; a few, e.g. Blenheim Orange apples, are best stored at normal air (21 per cent) pressures. It may be that the effect of reduced oxygen concentration works via its effect on ethylene production rather than directly on respiration. In the absence of oxygen, ethylene is not formed at all; but anaerobic respiration continues and the fruit soon breaks down.

It has also been found that raising the CO_2-concentration to 5–10 per cent may also markedly retard respiration rate and ripening, apparently because such concentrations retard the rate of oxidation of succinate to fumarate. They also inhibit PEP carboxylase, which normally fixes CO_2 and so replenishes malate and other acids lost from the citric acid cycle. In lemons this appears not to happen, and their rates of respiration and ripening are actually accelerated by about 10 per cent CO_2.

In commercial 'gas storage' of fruits, the three factors: temperature and oxygen and carbon dioxide concentrations are adjusted to the requirement of the particular fruit.

The idea that the specific inhibition of ethylene production might add a further method of control has led to investigations of ethylene formation in sundry plant tissues. Methionine, $CH_3SCH_2CH_2CHNH_2COOH$, and some other compounds have been identified as likely starting points, but not enough is yet known about intermediate stages to suggest methods of inhibiting them.

Further Reading

Bartley, W., Birt, L. M. and Banks, P. (1968) *The Biochemistry of the tissues*, Chapter 12, The Blood. Wiley, New York and London.

Biale, J. B. and Young, R. E. (1962) The biochemistry of fruit maturation. *Endeavour*, **21**, 164–74.

Burton, A. C. (1969) The mechanics of the red cell in relation to its carrier function. *In Circulatory and Respiratory Mass Transport*. Churchill, London.

James, W. O. (1953) *Plant Respiration*, Chapter 7, Oxygen effects. University Press, Oxford.

Mapson, L. W. (1970) Biosynthesis of ethylene and the ripening of fruit. *Endeavour*, **29**, 29–33.

Marchant, R. and Smith, D. G. (1968) Membranous Structures in Yeast. *Biol. Rev.*, **43**, 459–80.

Roodyn, D. B. and Wilkie, D. (1968) *The Biogenesis of Mitochondria*. Methuen, London.

8 *The Regulation of Respiration*

CELL metabolism is a tightly co-ordinated whole and to a very large extent self-regulating. It responds, however, to changes in the surroundings; the metabolism of unicells varies with the external environment, for example with its aeration; that of tissue cells with the internal environment, e.g. with the sugar content of the blood stream, and in plants there is an environment internal to the cell itself, the vacuole, with whose contents the cytoplasmic metabolism is equilibrated. A response to change in these environments is not limited to a single, simple reaction; the integration of metabolic reactions is so intimate that change in any single stage or equilibrium sets up changes in others to greater or lesser degrees.

Respiration, even in the wide sense that we have taken for it, is not segregated from the remainder of metabolism. The name is one we use for convenience and the cell does not oblige us by setting any sort of limitation round it. The agents of all sections of metabolism are essentially similar, and such opposed tendencies as respiration and synthesis, even including the photosynthesis of green cells, utilise many similar means and reactions.

Enzymes and membranes

The question to be considered is how this intricate medley is organised into an orderly system with powers of adjustment to the environment. At least so far as respiration is concerned, the tools at the cell's disposal are mainly of two sorts, enzymes and membranes. The two are not sharply divided, since many, perhaps most, if not all, enzymes are attached to cell membranes of one sort or another; the enzyme molecules may even be an integral part of the membrane's structural layer. Also the 'permanence' of both enzyme molecules and the membranous arrangement of given sets of molecules is doubtful and probably variable.

In this chapter we shall be mainly concerned with the enzymes. Enzymes are usually first thought of as accelerators of biochemical reactions; but as they exist and function in living cells they are much more than only this. Two properties are specially significant; their very high degree of selectivity and their sensitiveness to a wide range of controls. Both depend on their protein structure. The enzyme molecule is usually very much larger than the substrate and product molecules, which make contact with the enzyme at one or more restricted points on its surface, distinguished as the prosthetic group. But the activity of the enzyme is affected by the state of other parts of its molecule, a familiar example being the thiol,-SH, group, whose redox state is often of primary importance, even though it may not be directly concerned with the substance acted upon. As a result, many factors of the cell composition affect the activities of the respiratory enzymes, causing them to vary from full activity to zero, i.e. to a temporary or complete suppression.

Enzyme specificity

In the first place, the high specificity of enzymes is obviously a factor tending towards orderliness in the progress of reaction sequences; the enzymes do not react with the wrong substrates. Moreover, cells possess series of enzymes whose specificities are such that the product of one is the specific substrate of the next, thus leading to a controlled series of reactions in a given 'meaningful' direction. Many respiratory enzymes have two substrates, for example they may transfer hydrogen (or electrons) from a donor to acceptor, and they are usually highly specific to both. Hydrogen may be transferred from glyceraldehyde-3-phosphate to either NAD or NADP, but the two transfers are carried out by two different enzymes only the first of which, that using NAD as acceptor, probably takes part in respiration.

Nevertheless, many respiratory substances are not attacked by only one enzyme; NAD, acetyl CoA, pyruvic, isocitric and oxaloacetic acids may each be substrates for several. There may therefore be competition for any one of them among the enzymes concerned. Pyruvic acid, for example, may be acted on by an aminase, transaminases, decarboxylases, lactic dehydrogenase, pyruvic carboxylase and the malic enzyme (Figure 34). When lactic dehydrogenase is getting the lion's share, the main product will be lactic acid; but when the decarboxylases are most active the ultimate product will be the citric cycle acids or ethanol, depending on whether conditions are aerobic or anaerobic. The aminase and transaminases, lead in the direction of protein synthesis and the malic enzyme and pyruvic carboxylase towards resynthesis of carbohydrate. The enzyme equipment of the cell thus influences not only the rate at which metabolism proceeds, but the direction which it takes. It is not to be

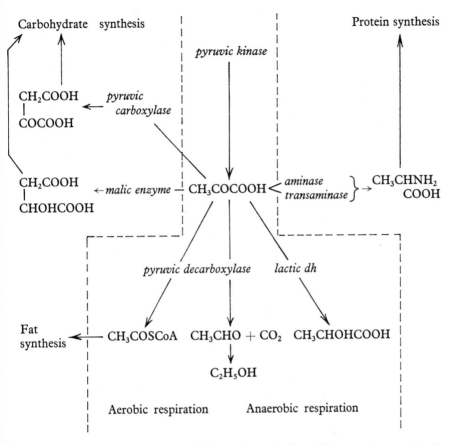

Figure 34 Reactions of pyruvic acid in cell metabolism. The reactions enclosed within the broken lines fall within the main streams of respiration. Enzymes are shown in italics.

supposed, however, that in all living cells direct competition for pyruvic acid is going on all the time among all these enzymes as it would in a homogeneous mixture; both the effective quantity of any enzyme and its effective 'affinity' for the substrate varies from one cell to another and with varying conditions in the same cell. Further, it is highly probable that some substrates such as NAD, are to some extent compartmentalised within the cell and, so to speak, allocated to particular enzymes.

Cofactors

All enzymes are affected to greater or lesser extents by the acidity of the medium in which they are at work. The average acidity of living cells is fairly constant, in animal cells around pH 7 and in plant cells around pH 6. Very little is known, however, about the acidity at particular positions where given enzymes may be located, or the degree to which it may be varied. Many other ions also affect enzymes more specifically; e.g. K^+ and Mg^{2+} are both needed by pyruvic kinase which catalyses the reaction.

78)   phosphoenolpyruvate $+$ ADP $\overset{\textit{pyruvic kinase}}{\rightleftharpoons}$ pyruvic acid $+$ ATP

Mg^{2+} is also required by pyruvic decarboxylase (cf., p. 45). Fe^{2+} is required for the full activity of aconitase and its removal or oxidation causes a block in the citric acid cycle (cf. , p. 64). Very many other such examples are known; but the method of their operation is often obscure. The term cofactor is used to cover a wide range of substances which may have a marked effect upon or be essential for an enzyme's activity. With increasing knowledge of the mechanisms of metabolic sequences, indefinite cofactors tend to pass into more precise categories. Thus the cofactors of the oxidative decarboxylation of pyruvic acid, diphosphothiamine and lipoic acid, are probably catalysts reacting directly with the intermediate compounds along the sequence (p. 57). Although the substances that may influence enzyme activity in a cell are thus very numerous, little is yet known about how or to what extent they cause adjustments in metabolic rates and directions by variations in their own concentrations at effective points.

Metabolites; enzyme repression

Perhaps the most interesting and significant substances in this respect are the metabolites themselves. These are obviously liable to fluctuations of concentration at the enzyme surfaces and it turns out that they may be both accelerators and inhibitors. These actions are quite independent of their mass action as primary reactants and affect a different part of the enzyme. It has long been known that many enzymes are unstable, or may not be produced at all in the complete absence of the appropriate substrate. Such adaptive enzymes are particularly familiar in unicellular microorganisms which may be 'trained' to accept unusual nutrients, developing appropriate enzymes in the process. Similar changes can also occur in tissue cells. For example α-amylase activity, the ability to hydrolyse α-1, 4 linkages in moderately long chain glucans, was found in germinating barley tissues only so long as the suitable substrates were also being formed from starch. The amount of enzyme rose and fell with the concentration of the substrate. During starvation, the enzymes which convert glucose to glucose-1-phosphate and phosphogluconic acid diminish rapidly in the liver, whereas glucose-6-phosphatase, which releases glucose to the blood stream, is not lost but may even increase. Conversely, livers of animals fed after a period of starvation show an increased formation of the enzymes of the pentose phosphate pathways. Dramatic changes of enzyme content, for example in the amounts of glucose-6-phosphate dehydrogenase and malic enzyme occur in rat liver cells at the time of weaning, i.e. when the diet becomes less rich in fats and lipid synthesis is promoted in place of breakdown. Since all the cells of a given organism have the same genetic constitution and therefore the same potentialities, such long-term effects, controlling the rates at which enzymes are synthesised, are referred to as enzyme repression and often a whole sequence may be involved simultaneously.

Effector metabolites

Regulatory action of metabolites is not limited to long-term effects or to the enzyme of which they may be a substrate or product; indeed more frequently they influence an enzyme catalysing some antecedent stage of the sequence, so setting up a 'feedback mechanism' that tends to restrict or accelerate their own formation according to whether their effect is inhibitory or the reverse. An example is given by the action of oxaloacetate, the concentration of which is normally kept very low in cells. This low concentration is required for the continuance of the citric acid cycle; but if the concentration rose it would become a potent inhibitor of succinic dehydrogenase which, through fumaric and malic acids, leads to the formation of the oxaloacetic acid itself. The rate of production of oxaloacetic would thus be automatically cut down until the removal reactions had again reduced its concentration.

Regulations of this type may be most significant at points where two or more reaction sequences diverge from a single metabolite. A notable example for respiration is afforded by glucose-6-phosphate. During normal respiration glucose-6-phosphate is being formed from glucose-1-phosphate, derived from glycogen, starch or other similar reserve. It has three main lines of utilisation indicated in Figure 35. It may be hydrolysed by glucose-6-phosphatase, and the glucose removed to other cells; it may be converted by phosphohexoisomerase to fructose-6-phosphate and so enter glycolysis; or it may be oxidised by its dehydrogenase to 6-phosphogluconic acid and so enter the pentose phosphate pathway. The hydrolysis is subject to inhibition by its own products, and the isomerase is subject to inhibition by 6-phosphogluconic acid and other products of the pentose phosphate pathway. Even if complete this would not necessarily lead to the total suppression of glycolysis when the PPP is active; because the pathway has fructose-6-phosphate and glyceraldehyde-3-phosphate among its products, and they are also intermediates in glycolysis.

Inhibitions and activations by effector metabolites, whether acting on their own or some preceding enzyme, usually occur by a combination with the enzyme at a site different from the prosthetic enzyme site and cause a conformational, or 'allosteric' change in the enzyme molecule. Enzymes susceptible to such modifications are frequently found to consist of several isoenzymes; lactic dehydrogenase has, for example been electrophoretically separated into five. These consist of a 'muscle' type, M, a 'heart' type, H, and the three 'hybrids', M_1H_3, M_2H_2 and M_3H_1. Particular cells may contain predominantly M or H or mixtures of all five. The mixture may change with stage of development. Thus rat foetal cells 9 days after fertilisation contain M with some M_3H_1

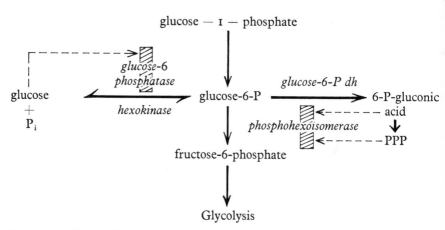

Figure 35   Metabolic reactions of glucose–6–phosphate. Enzymes in italics. Inhibitions indicated by broken-line arrows and cross bars.

whereas the adult heart cells contain mainly H and M_1H_3. In other species, including man, the lactic dehydrogenase of heart cells is predominantly of the H type from the beginning. The M type is stated to be more widely affected by metabolic inhibition than the H and to be allosteric; but in one important respect the H type is the more sensitive. It is inhibited by excess pyruvic acid whereas the M type is not. It has been suggested that this favours the accumulation of lactic acid in skeletal muscle, rich in the M form, and helps to retard it in heart muscle, with its predominance of the H isoenzyme. It is even said that birds which fly only for short and intermittent periods have M type enzyme in their breast muscle, whereas those capable of long, sustained flights have a high percentage of H type. The H isoenzyme is also more efficient than the M type in oxidising lactic acid at low concentrations and heart muscle is particularly efficient in oxidising the resulting pyruvic acid owing to the abundance of its mitochondria. Although the heart receives lactic acid in the venous blood stream it is thus a net oxidiser of it; whereas skeletal muscle may be a net producer.

The ratio [ATP]/[ADP] [P_i]

This is perhaps the most widely significant of the built-in regulators of respiratory rates. In the cell, respiratory oxidations whether aerobic or anaerobic are obligately bound with phosphorylation; i.e. they require free ADP and P_i to proceed. The chemical mechanism of this linkage in the anaerobic oxidation of glyceraldehyde-3-phosphate has already been described on p. 35; the exact nature of the tie-up during the oxidative stages of respiration is still open to conjecture; but, in either case, free inorganic phosphate is a requirement for the reactions to proceed. This was demonstrated by Harden and Young as long ago as 1905 for the glycolysis in yeast fermentation. Aerobically it has been shown to be needed in, for example, the respiration of liver mitochondria and barley embryos. Either [P_i] or [ADP] may be rate-limiting; in living cells probably both may play a rôle; e.g., in complete barley seedlings, reserves of inorganic phosphate are high enough to prevent its concentration dropping to limiting values. In embryos isolated from their supply tissues, but still capable of rapid respiration, inorganic phosphate may become the rate-limiting factor (Figure 36).

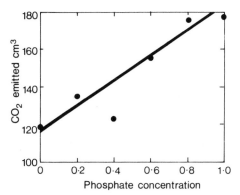

Figure 36 CO_2 emitted by 80 barley embryos germinated on culture solution containing 4 per cent sucrose. Phosphate concentrations are given as fractions of the concentration in a normal culture solution and CO_2 is summed from the second to sixth days of germination inclusive.

The total concentration of ADP + ATP in cells is always small (perhaps around 10 mM) and mature cells appear to have little ability to synthesise either *de novo*. It follows that oxidations, which are coupled with the conversion of ADP + P_i to ATP rapidly tend to lock up the nucleotide in the phosphorylated form and so slow down the rate of phosphorylation and with it the coupled oxidation itself. Respiration rates are therefore strongly influenced by reactions which release ADP + P_i from ATP and these include the great majority of metabolic syntheses and the utilisation of energy in muscular or flagellar activity. In short, the amount of respiration tends to be regulated by the amount of utilisation, which implies an economical expenditure of the energy and the reserves from which it derives.

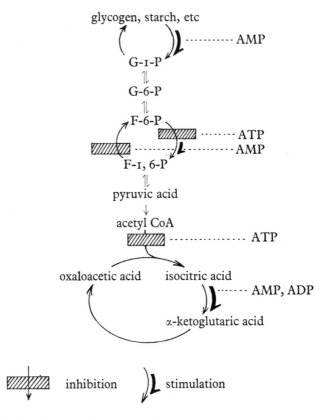

Figure 37 Adenosine phosphates as respiratory regulators.

ADP and ATP are involved in some of the respiratory reactions as substrates as, for example, during the phosphorylatory oxidation of glyceraldehyde-3-phosphate. But both these nucleotides and also AMP have been shown to act also as effectors controlling the affinity of several respiratory enzymes for their substrates. Figure 37 shows their major points of operation as inhibitors or stimulators. Yeast isocitric dehydrogenase is stimulated by AMP, which increases its affinity for isocitric acid; this applies to the enzyme in carefully isolated mitochondria. In animal cells the AMP is replaced by ADP as effector. Conversely, ATP strongly reduces the affinity of citric synthetase (condensing enzyme) for acetyl CoA. When ATP concentration is high, the citrate cycle will therefore be competing weakly for acetyl CoA. But when ATP concentration is high ADP (or AMP) concentration is always low, and the removal of isocitrate and therefore of citrate will be correspondingly retarded in the citrate cycle. The accumulated citrate (15 times that of isocitrate at equilibrium) acts as a positive effector of the acetyl CoA carboxylase, the enzyme that initiates fat synthesis. A high ATP/ADP ratio will therefore tend to favour fat synthesis at the expense of the oxidative cycle, until the energy-rich ATP is reduced again.

Hormones

The metabolic rates of cells forming parts of large organisms may be influenced by 'cofactors' which reach them from other cells of quite different types. Such hormones may be slow acting, as for example the sex hormones, which control the onset of puberty; or they may be fast acting like injected insulin which causes a rapid drop in the concentration of sugar in the blood. Slow-acting hormones are generally supposed to cause the synthesis of new enzymes and fast-acting ones the activation of enzymes already existing. Since metabolic reactions are so closely interknit it is a matter of great difficulty to decide the primary point at which an entering hormone affects the cell's economy, and which effects are secondary. Moreover the mechanism of reaction may be intrinsically complex. A single intensively studied example may be mentioned in relation to respiration. Adrenalin,

$$\text{HO} \diagdown \bigcirc \text{CHOHCH}_2\text{NH}_2\text{CH}_3$$
$$\text{HO} \diagup$$

is formed in cells of the adrenal cortex and accumulates in microbodies containing also protein and ATP. When transferred to liver or muscle cells adrenalin causes the breakdown of glycogen to glucose-1-phosphate which is catalysed by phosphorylase. Adrenalin does not, however, act directly upon the phosphorylase. Its immediate reaction appears to be upon an enzyme, adenyl cyclase, which produces cyclic-3',5'-AMP from ATP. The cyclic AMP in turn activates a phosphorylase kinase and it is this that finally converts phosphorylase *b* into the more active phosphorylase *a* form. The increased formation of glucose-1-phosphate that results increases both the respiration of the liver cells and the passage of sugar into the blood stream.

It has been possible to describe only a few examples of the different built-in controls that affect the rate and direction of respiratory events in cells. They have been chosen from the more generalised types of cell and high degrees of specialisation are often accompanied by specialised controls. It will be clear too, that the great versatility of enzymes not only affects events within a single sequence of reactions; but is, on the contrary, at its most important at the cross-roads of metabolism, where catabolisms and syntheses share common reactants such as glucose-6-phosphate, pyruvic acid, oxaloacetic acid or acetyl-CoA. On this account enzymes have been referred to as sensors of the intracellular environment translating its many variations into changes of catalytic activity.

Further Reading

Atkinson, D. E. (1965) Biological feedback control at the molecular level. *Science*, **150**, 851–7.
Bartley, W., Birt, L. M. and Banks, P. (1968) *The Biochemistry of the Tissues*, Section 8. Wiley, New York.
Boulter, D., Laycock, M. V., Ramshaw, J. and Thompson, E. W. (1970) Amino acid sequence studies of plant cytochrome *c. In Phytochemical Phylogeny*. Academic Press, London.
Fine, I. H., Kaplan, N. O. and Kuftince, D. (1963) Developmental changes of mammalian lactic dehydrogenases. *Biochem.*, **2**, 116–21.
Frisch, L. (editor) (1961) *Cellular Regulatory Mechanisms*. Cold Spring Harbour Symp. Quant. Biol., New York.
Kaplan, N. O. (1963) Multiple forms of enzymes. *Bacteriol. Rev.*, **27**, 155–69.
Loewy, A. G. and Siekivitz, P. (1970) *Cell Structure and Function*. 2nd ed., Chapter 10. Holt, Rinehart and Winston, London.
Paigen, K. and Williams, B. (1970) Catabolic repression and other control mechanisms in carbohydrate utilisation. *Adv. in Microbial Physiol.*, **4**, 251–324.
Stadtman, E. R. (1966) Allosteric regulation of enzyme activity. *Adv. in Enzymol.*, **28**, 41–154.

9   *Respiration and Fine Structure*

MANY of the processes tending to regulate respiration described in the last chapter might occur in a more or less homogeneous system. In addition to such controls it is evident that the suitable arrangement of groups of enzymes responsible for sequential reactions might well increase their over-all efficiency; conversely spatial separation of the groups might prevent destructive interactions. Grouping of enzymes may occur in a variety of ways, as also their separation. Cell membranes (cf., p. 11) may serve to carry enzymes in fixed positions, for example cytochromes on mitochondrial walls, or to segregate them from the general cytoplasm (hydrolases in lysosomes); they may also control the rates of supply of substrates and the removal of products by the degree of their permeability to them.

Cell fractions

Present knowledge of the intracellular localisation of enzymes depends mainly upon the separation of subcellular particles from disintegrated cells and to a lesser extent on cytochemical methods observing the responses of unbroken cells to applied reagents. The separation of different categories of particles depends almost wholly on the centrifugation of the cell brei in bland suspending media, usually aqueous, but occasionally for special purposes organic. The media are often so arranged as to provide a continuous or discontinuous density gradient. The fractions obtained are monitored by electronmicrographs to correlate them with organelles of the unbroken cell. The practice of identifying fractions by 'marker' enzymes considered to be specific for a given fraction led to considerable confusion in the early stages. Such a method can only be used when much is already known about the investigated systems.

The endoplasmic reticulum, the membrane system which ramifies through the cytoplasm of eucaryotic cells, breaks up during extraction and prolonged centrifugation above 10 000 $\times$ g into fragments which may have attached ribosomes. This precipitate is referred to as the microsomal fraction and under the electron microscope may be seen to have formed vesicles and other arte-facts during isolation.

One of the most complex parts of the cell, regarded as a site of metabolic action, is the 'soluble cytoplasm' left over when all the precipitable particles have been removed by prolonged centrifuging above 100 000 $\times$ g. The thought naturally arises that some enzymes in this fraction have been dis-solved from the surfaces or fragments of organised particles during the pro-longed treatment. Careful cross comparisons with other fractions and with the total homogenate, together with the discreet use of non-aqueous media have now done a good deal to control this source of error.

The soluble cytoplasm: glycolysis

The core of respiratory metabolism in almost all types of cells is the glycolysis of hexose phosphates to pyruvic acid. It is an interesting fact that this primi-tive, or at least general, respiration appears to occur in the soluble cytoplasm; i.e. with a minimum of dependence on membranes. The entire process can occur in clarified cell saps; and the enzymes that catalyse its successive re-actions remain almost wholly in the supernatant after all sub-cellular particles have been spun down. This has been shown for such diverse cells as those of brain, liver, pituitary gland, muscle, various tumours, roots, leaves and yeast. The same is also true of the enzymes which complete the anaerobic respiration of pyruvic acid; lactic dehydrogenase in animal and some plant cells; pyruvic decarboxylase and alcohol dehydrogenase in plant cells (e.g. carrots, barley roots, chrysanthemum leaves) and yeast. The alcohol dehydrogenase of liver cells is also in the soluble phase.

When respiration begins with a glucose unit, the reactions which initiate it, catalysed for example by hexokinase and phosphofructokinase, are also known to be largely associated with the same cell phase. In liver cells the glucose is derived from glycogen, which accumulates as small particles in the cytoplasm (Figure 7, p. 27). The α-glucan phosphorylase that converts it to glucose-1-phosphate is also distributed through the cytosol. During centrifugation part of the phosphorylase comes down attached to the glycogen particles. The same appears to be true of the α-amylase that hydrolyses glycogen to glucose.

In plant cells the situation may be rather more complex. Instead of glycogen they store starch or sucrose. The former is always deposited in plastids which are separated from the general cytoplasm, usually by a double unit membrane. Each starch grain may be individually surrounded by a single membrane derived from the inner layer of the plastid envelope. Starch has a more com-plex molecular structure than glycogen with branching caused by occasional α-1,6 linkages in addition to the usual α-1,4 linkages building the length of the spiral chain. The phosphorylases that catalyse its formation and degradation to glucose-1-phosphate are more numerous, and it is uncertain whether the same enzymes catalyse the sequence in both directions. There are also amylases which, in association with maltase, hydrolyse starch to glucose. The earlier stages of starch breakdown occur within the plastids; but at what stage the products are released into the soluble cytoplasm is unknown. The β-amylase which assists the breakdown of starch stored in germinating peas is localised in small 'aleurone vacuoles', which also contain proteases and numerous other hydrolysing enzymes released at this time. This lysosomal action may be regarded as a special provision for the rapid metabolism of germination.

Sucrose is formed primarily in chloroplasts, but it is stored in the vacuoles of all plant cells. Although a sucrose phosphorylase, catalysing the equilibrium

79) $$\text{Sucrose} + P_i \rightleftharpoons \text{fructose} + \text{glucose-1-phosphate}$$

is known from bacteria, it does not appear to exist in plants. Sucrose formation appears rather to depend on a sucrose synthetase requiring UDPG (uridine diphosphate glucose) as an intermediate. A similar sucrose phosphate synthetase may give sucrose after hydrolysis of the sucrose phosphate. In tissues such as carrots, and also in some yeasts, sucrose is consumed in respiration. Whether it is mobilised by a synthetase acting in reverse or by invertase and in what part of the cell is unknown. It is perhaps significant that the system synthesising sucrose phosphate apparently occurs both in the soluble cytoplasm itself and in chloroplasts. Sucrose itself can apparently pass the chloroplast membranes.

The pentose phosphate pathway

The soluble cytoplasm also contains the enzymes of the pentose phosphate pathway and at least some of the NADP oxidase activity, i.e. the capacity to transfer hydrogen from $NADPH_2$ to atmospheric oxygen. This is not carried out by mitochondria, and may be associated with soluble oxidases like the catechol and ascorbic acid oxidases. It is apparently not linked with any ATP formation (cf., Chapter 6). Enzymes of the pentose phosphate pathway have been attributed to the soluble cytoplasm in liver, kidney, sundry plant tissues and yeast. In procaryotic cells, i.e. those without highly developed internal membrane systems, they may be located in the inner membraneless protoplasm as, for example, the glucose-6-phosphate dehydrogenase of *Staphylococcus aureus* (Table 1, p. 5). Adult red blood cells, which have lost their nuclei and mitochondria retain an active PPP upon which they depend for the integrity of their bounding membrane (p. 71).

Structure within the soluble cytoplasm

Glycolysis, PPP activity, and NADP oxidation and reduction are by no means the only activities that appear to take place in soluble cytoplasm; and it might well seem that their regulation was little affected by considerations of fine structure. It must, however, be remembered that we know very little about the physical nature of this fraction. The methods of separating it from the particulate and membraneous fractions involve its considerable dilution with liquid media. It is rich in proteins, usually containing around 70 per cent of the total amount in the cell, and probably often has gel structure. Even in a proteinaceous sol, functional but still soluble aggregates of enzymes might well be formed, such as those built by the A and B type specificity of NAD (p. 38).

The ramifications of the endoplasmic reticulum are exceedingly complex and it is not yet possible to say whether it does or does not divide the soluble cytoplasm into segregated compartments. Even apart from a total isolation, the formation of pockets might exercise controls on the direction of reaction. Compartmentation within the soluble cytoplasm has been invoked to explain the fact that insulin does not affect the glycolysis of glycogen to lactic acid in rat diaphragm cells, but does stimulate its synthesis from glucose. On the other hand enzymes involved in the glycogen $\rightleftharpoons$ glucose equilibrium, such as phosphofructokinase and fructose-1,6-diphosphatase are known to be sensitive to effector metabolites and it might be through them that the insulin effect operates.

The mitochondria

The whole apparatus of oxidative phosphorylation, including the citric acid cycle, the hydrogen and electron transport chain and the phosphorylations associated with it have now been shown, in a very wide range of cell types, to occur in the mitochondria. In cells that are capable of alternative aerobic or anaerobic states, notably the yeasts, the capacity for aerobic respiration and the condition of the mitochondria are closely allied (p. 104). Procaryotic cells without complex internal membrane systems, but which are nevertheless capable of aerobic respiration, carry their cytochromes and associated enzymes on the plasma membrane (Table 2, p. 8). Similarly, in the blue-green algae, the sites of aerobic respiration may lie in the outer layers of the protoplasm (Plate II) on a system of unenclosed lamellae which they share with photosynthesis. The general structure of mitochondria has been described in Chapter 1 (p. 13).

The location of respiratory enzymes

A vast amount of work has been devoted to the attempt to locate the sites of the oxidations and phosphorylations more minutely, but so far with only very moderate success. It may, however, be taken as a starting point that all the respiratory processes are based upon the mitochondrial membranes and probably more particularly upon the inner membrane and the cristae or microvilli that are continuous with it. It is noticeable that high respiratory activity is associated with a more abundant development of this particular surface in cells of the most varied types. Not only do cells, such as heart muscle cells, with a constant high energy output have particularly abundant mitochondria with highly developed cristae; but cells with notable changes of aerobic respiration rate tend to show corresponding changes of mitochondrial structure. An interesting example is given by *Arum* spadix. During the unfolding of the spathe (Figure 27, p. 89) the aerobic respiration rate temporarily attains a very high peak. It has been recorded that temporary increases of microvillar surface can be observed at the same time.

Although the oxidative processes appear to be based on the membrane structure of the mitochondria rather than in the inner or outer spaces, the firmness with which the constituent enzymes and carriers are attached to the membranes varies greatly. When the mitochondrial walls are disintegrated mechanically or by ultrasonic vibration, they break up into particles which retain the power to oxidise $NADH_2$ or succinic acid through the electron transport system (Figure 23, p. 77), and which contain the constituents of the system in more or less stoichiometric amounts. Heart muscle results are shown in Table 6. Rather similar results were obtained with particles from *Azotobac-*

TABLE 6

Approximate molar or atomic ratios of redox components in electron transport particles of heart mitochondria (After Green and others)

NAD	10
NADP	2
$NADH_2$ dehydrogenase (flavoprotein)	1
Succinic dehydrogenase (flavoprotein)	1
CoQ	15
Non-haem iron	18
Cytochrome b	2
Cytochrome $c + c_1$	2
Cytochrome a	2
Copper	6

Lipid content about 20 per cent.

ter vinelandii; but plant mitochondria may give a considerable excess of cytochrome *b*. It has been estimated that there may be 17 000–18 000 such assemblies in a heart mitochondrion and probably fewer in a liver one with its smaller cristae.

Electron transport particles

Many of the constituent enzymes, notably cytochrome $a + a_3$, cannot readily be extracted from these particles, though they can be shown, spectrophotometrically, for example, to be still active within them. Their release involves a violent chemical disruption of the lipoprotein membrane with bile salts, detergents and the like. So far no reconstitution of the electron transport mechanism, let alone any rebuilding of its associated phosphorylations, has been achieved by reassembling its isolated members. The idea of a closely knit electron transport particle with a special structure and embedded in the mitochondrial wall seems well founded.

The attempt to get a closer knowledge of its structure, i.e. of the way in which the individual enzymes and carriers are arranged within it, has proved to be fraught with many difficulties. At present the use of various extractants and the careful examination of the behaviour of their products towards the known inhibitors of the electron transport system has suggested that the particle may include four assemblies, each of which has been isolated more or less free from components of the others. The four complexes and their supposed functional connections through CoQ and cytochrome *c* are illustrated in Figure 38.

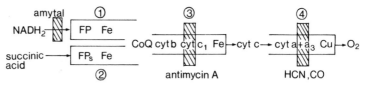

Figure 38 Diagram to illustrate the proposed system of electron-transport complexes ① to ④. Cross bars indicate points of action of inhibitors.

The total electron transport particle includes, besides the enzyme carriers shown in Figure 23, lipoproteins, i.e. 'complexes of lipid and protein which behave as *bona fide* molecules' and which include a bewildering variety of lipids. There are also non-haem iron and copper. The composition of the four constituent complexes as isolated from ox heart are briefly as follows. Complex (1) of Figure 38 contains the flavoprotein $NADH_2$ dehydrogenase and about 16 atoms of non-haem iron for each molecule of flavin; also some CoQ and traces of cytochromes *b* and c_1 possibly as impurities. There is about 20 per cent of lipids. Reduction of CoQ analogues (Q_1 and Q_2) is inhibited by amytal, and, owing probably to the presence of the 'impurities', some reduction of cytochrome c can occur. Complex (2) contains succinic dehydrogenase, non-haem iron, lipid, some CoQ and traces of cytochromes b and c_1. Its reductive activity in the presence of succinic acid is not inhibited by amytal. Complex (3) contains cytochrome b, cytochrome c_1 non-haem iron, lipid and a little flavin. It is completely inhibited by antimycin A. Complex (4) contains cytochrome $a + a_3$ and copper.

It will be seen that the complexes as isolated are not completely pure though consisting predominantly of given segments of the electron transport chain. When remixed together with cytochrome c they reconstitute the entire chain. If complex (1) is omitted only succinic acid is oxidised; if complex (1) is added and complex (2) omitted, $NADH_2$ is oxidised and not succinic acid. The status of CoQ is ambiguous; it is always found associated to some extent with complexes (1) and (2) and added Q_{10} is not reduced though its analogues are.

Membrane-requirement of oxidative phosphorylation

Although the reassembled complexes can unite to form a complete $NADH_2$ or succinic oxidase system, they do not bring about any phosphorylation. This has not yet been achieved in the absence of a membrane system and, in spite of the most extensive searches, no intermediate high energy compounds have been certainly identified which might link electron transport with ATP formation. This is in sharp contrast with the glycolytic ATP formation as described on p. 35. The hunt goes on, but in the meantime there remains the possibility that no such compounds exist and that the generation of labile ATP bonds depends upon a charge separation across a membrane in the manner described on p. 86. As a result of experiments with rat liver mitochondria it has been suggested that cytochromes b and $a + a_3$ are sited on the inner surface of the cristae (i.e. towards the matrix) and cytochromes c and c_1 on the outer. The generation of ATP would therefore be due to a charge separation across the membranes of the cristae, presumed impermeable to H^+ and OH^- ions, and the anisotropic arrangement of at least the late members of the chain transferring electrons to oxygen; i.e. to a physical and structural cause rather than a purely chemical one.

The site of enzymes of the citric acid cycle

The enzymes of the citric acid cycle are also found in the mitochondria; though similar, but not necessarily identical, enzymes are also found in the soluble cytoplasm. The soluble and mitochondrial malic dehydrogenases of ox heart, for example, have been shown to have very different amino acid compositions. Although succinic dehydrogenase is usually regarded as the mitochondrial dehydrogenase *par excellence*, soluble variants of the enzyme occur in yeast grown anaerobically and in its 'petite' mutants; also in chicken embryos and rat mammary glands. On the other hand it has been identified firmly attached to the mitochondria of a vast range of animal and plant cells.

The pyruvic and α-ketoglutaric dehydrogenase systems appear to be associated exclusively with mitochondria. They are themselves organised enzyme systems of considerable complexity (cf., p. 56) and high particle weight. The other enzymes of the citric cycle do not appear to form such complexes, but are found more loosely associated.

Although malic dehydrogenase (NAD-specific) has been found in the soluble cytoplasm of a wide range of tissues, from rat brain to locust muscle and pea epicotyls, a proportion of it always remains associated with the mitochondria. In those of pea epicotyls it is associated with the wall fraction and is extracted by aqueous solvents with about the same ease as cytochrome *c*. In considering the relation of malate reduction with the electron transport chain, it is interesting to remember that malic dehydrogenase has A-type specificity for NAD whereas $NADH_2$ dehydrogenase, and therefore complex (1) of the electron transport system, has B-type. A close link can thus be made between this step of the citric acid cycle and the electron transport chain.

Isocitric dehydrogenase (NAD-specific) appears to be limited to mitochondria; but a NADP-specific enzyme occurs in both mitochondria and supernatant. Fumarase is also, at least in part, associated with the mitochondria in plant and animal cells, and the same is true of aconitase. It has been shown that the mitochondrial and supernatant aconitases of rat liver have very different pH responses, and may therefore be different enzymes. Mitochondria isolated from pea epicotyls do not lose aconitase when driven centrifugally through sucrose solutions of varying strengths. When they are subsequently frozen and rethawed they yield a matrix sap with high aconitase activity and the walls retain none.

With the exception of aconitase, it appears that all the enzymes of electron transport and the citric acid cycle are attached with varying degrees of firmness to the walls and that the aconitase is concentrated in a chamber inside them. As already explained the inner wall, including its cristae and microvillar extensions, seems likely to be the more important of the two layers. Experimental and electron-microscopical attempts to decide how the respiratory assemblies are arranged upon it and how they are related to its structural protein have led to results of great complexity which are still very controversial.

The reduction of mitochondrial NAD

The NAD contained in mitochondria is liable to reduction by four different dehydrogenases of the citric cycle, (pyruvic, α-ketoglutaric, isocitric and malic) by glutamic dehydrogenase and, when fatty acid oxidation is occurring, by the dehydrogenases of β-hydroxyacyl-CoA compounds. It may also be liable to reduction by isocitrate through NADP (Figure 39). With the possible exception of the last, there is no evidence that the NAD is compartmentalised in any way that would structurally control a 'fair share' of the oxidising power to each of these sources. It appears rather that they are in competition for the

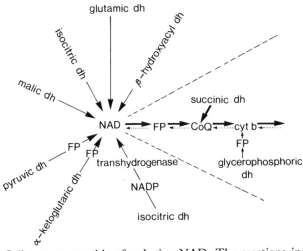

Figure 39 Cell systems capable of reducing NAD. The reactions included within broken lines to the right may become reducers under special conditions as shown by the dotted arrows and described in the text.

NAD and the entry it affords to the electron transport chain. Furthermore, the NAD branch is itself in competition as hydrogen donor with the α-glycerophosphate 'shuttle service' to the soluble phase (p. 128) and with the succinic acid branch (Figure 39). It is claimed that the system may be so saturated with succinic acid that the entire NAD branch may be jammed. It is even known that, given an excess of ATP (i.e. high [ATP]/[ADP] [P_i]), and a terminal block due to lack of oxygen or a cytochrome $a + a_3$ inhibitor, electrons may be 'forced back' along the entire chain and NAD be reduced instead of $NADH_2$ being oxidised. It would appear that the orderly feeding of the electrons into the unified chain must depend upon dynamic regulators and feed backs of the kind discussed in the last section as well as upon purely structural features.

The mitochondrion certainly seems to be the main site of respiratory oxidation in most eucaryotic cells. It is, however, by no means the only place where direct reaction with oxygen goes on. Oxidases of one sort or another may be found in almost all parts of aerobic cells. The best known of these are described in Chapter 6, which shows that their relations with respiration are often obscure or of a specifically limited kind, rarely, if ever, including any formation of ATP.

The oxidation of soluble $NADH_2$

The $NADH_2$ involved in the respiration chain of Figures 23 and 39, like most of the other components, is firmly attached to the mitochondrial structure and, on the evidence of labelling experiments, further $NADH_2$ is not readily taken up from outside. On the other hand, the $NADH_2$ produced when glyceraldehyde-3-phosphate is oxidised in glycolysis (p. 35) is located in the 'soluble cytoplasm' outside the mitochondria, and the reaction continues aerobically when pyruvic acid is no longer available as oxidant. The glyceraldehyde-3-phosphate exists in an equilibrium that strongly favours its isomer dihydroxyacetone phosphate. In the presence of the appropriate dehydrogenase, the latter oxidises $NADH_2$ yielding α-glycerophosphate and reoxidised NAD

80)
$$
\begin{array}{ccc}
\text{CH}_2\text{OH} & & \text{CH}_2\text{OH} \\
| & \xrightarrow{\;L\text{-}\alpha\text{-}glycerophosphate\ dh\;} & | \\
\text{CO} + \text{NADH}_2 & \rightleftharpoons & \text{CHOH} + \text{NAD} \\
| & & | \\
\text{CH}_2\text{O}\,\textcircled{P} & & \text{CH}_2\text{O}\,\textcircled{P}
\end{array}
$$

      dihydroxyacetone phosphate   α-glycerophosphate

Although mitochondria are impermeable to $NADH_2$ they are permeable to α-glycerophosphate. The mitochondria of many cells, for example in brain, liver, muscle, kidney and fatty seeds, have been shown to contain a second α-glycerophosphate dehydrogenase of a different type which transfers reducing units to cytochrome *c*, probably by way of a flavoprotein and cytochrome *b*. The reformed dihydroxyacetone phosphate returns to the soluble cytoplasm and so establishes a sort of substrate shuttle service by means of which extramitochondrial $NADH_2$ may be continuously reoxidised. The occurrence of such a service has been established for the wing-muscle cells of some insects, and it has been proposed that it, or perhaps similar shuttles using other substrates like malate or β-hydroxybutyrate, may be widely distributed.

Glycerol may be formed during glycolysis (p. 46); but it is formed more abundantly whenever fats are being respired, notably in liver cells and the cells of fatty seeds during germination. If oxidation proceeds by way of an initial phosphorylation to α-glycerophosphate

81)
$$\begin{array}{c} CH_2OH \\ | \\ CHOH + ATP \\ | \\ CH_2OH \end{array} \xrightarrow[\text{glycerol kinase}]{} \begin{array}{c} CH_2OH \\ | \\ CHOH + ADP \\ | \\ CH_2O\text{\textcircled{P}} \end{array}$$

which is then oxidised by NAD to dihydroxyacetone phosphate, the reverse phase of the reaction 80) above, this may then enter the glycolysis series of reactions as glyceraldehyde-3-phosphate and be oxidised in the usual way via pyruvic acid, the TCA cycle and the electron transport chain. Evidence for this has been found in the cells of liver, peanuts and castor oil seeds; but in all these examples the bulk of the triose phosphate is synthesised back to sugars.

In some strains of *Trypanosoma* the same type of oxidation occurs; but other strains, although they are capable of oxidising the α-glycerophosphate quantitatively, do not contain any cytochromes. Their oxidation system is bound to particles, but the particles have not yet been shown to have mitochondrial structure.

It is probable that soluble oxidising systems for $NADH_2$ exist outside the mitochondria (cf., p. 97); but little is known about them at present.

The oxidation of $NADPH_2$

This dinucleotide is also found both in the mitochondria and the soluble cytoplasm, as well as in the chloroplasts of green cells. It also appears to be unable to pass from one phase to another. Glycolysis and the citric acid cycle give rise mainly to $NADH_2$; but mitochondria contain two isocitric dehydrogenases one of which reduces NAD and the other NADP. There is a similar pair of glutamic acid dehydrogenases. Mitochondria from liver, kidney, brain, etc. possess a tightly bound transhydrogenase which catalyses the shuttling of hydrogen atoms between NAD and NADP, and it is supposed that mitochondrial $NADPH_2$ is oxidised by the respiratory chain via NAD (Figure 39).

Perhaps the main respiratory generator of $NADPH_2$ is the pentose phosphate pathway (p. 71), and this occurs outside the mitochondria. Soluble systems are known which will oxidise $NADPH_2$ at the expense of oxygen, of which one of the best known is

82) $NADPH_2 \xrightarrow[]{\text{glutathione reductase}} \text{glutathione} \xrightarrow[\text{dehydroascorbic reductase}]{} \text{ascorbic acid} \xrightarrow[]{\text{ascorbic oxidase}} O_2$

All these components are widely distributed; but neither this nor any similar system has been shown to be operative *in vivo*. Nor are any known to couple with ATP-formation.

It seems likely that, even under aerobic conditions, much of the $NADPH_2$ generated inside cells by the pentose phosphate pathway is not oxidised in this sense at all. Instead it is reoxidised to NADP, when its reducing power is used internally, for example, in the carboxylation of pyruvic acid (p. 73) or the synthesis of fats (p. 71).

The microsomal fraction

This fraction is usually regarded as consisting of fragments of endoplasmic and other membranes, carrying ribosomes with them. There is evidence that the membrane fragments recovered from a wide variety of cells carry cyto-chrome reductases reacting with both $NADH_2$ and $NADPH_2$. Although they can be assayed with cytochrome c this is not found associated with the membranes; but, instead, there is cytochrome b_5 and, in plants, cytochrome b_3. Membrane-bound dehydrogenases that might reduce the oxidised NAD and NADP again have not commonly been reported; but reaction with those of the soluble protoplasm is not necessarily excluded. These systems are not however known to be linked with any ATP formation or to make a significant contri-bution to the energy resources of the cell.

Chloroplasts and respiration

Green cells are in the unique position of including within a single cell two such different techniques for the production of ATP as the photo-reductive and the thermal-oxidative; but although the starting points are so different the methods employed may, in fact, have many similarities. In the green cells of the higher plants the electron transport phases of respiration and photosyn-thesis appear to be rigidly separated into separate membrane-bounded orga-nelles, the mitochondria and chloroplasts respectively. Contrary to what is sometimes said, green cells usually have abundant mitochondria. The plasto-quinone and cytochrome b_6 and f that are bound to the chloroplast thylakoids, and which may afford part of its 'reversed' electron transport pathway, are similar to, though not identical with, the ubiquinone and cytochromes b and c of the mitochondria. The predominant dinucleotide at the end of the chain is NADP rather than NAD. Respiration is not inhibited by light, and it is not evident that the small differences in the chloroplast and mitochondrial consti-tuents would in themselves enable two separate electron streams to operate in opposite directions simultaneously. In this case, the spatial separation appears to be the effective factor.

The reducing power generated in the form of $NADPH_2$ in the chloroplasts may be used to reduce 3-phosphoglyceric acid to glyceraldehyde-3-phosphate.

This is the reverse of the glycolytic oxidation (p. 35); but with NADP replacing NAD. The triose phosphate dehydrogenase that reacts with NADP in the reaction

83) $CHO\ CHOH\ CH_2O\textcircled{P} + NADP + P_i \rightleftharpoons COO\textcircled{P}CHOH\ CH_2O\textcircled{P} + NADPH_2$

is situated in the stroma of the chloroplast and is not found in the soluble cytoplasm, though the NAD-specific enzyme operating in glycolysis is probably not limited to the soluble cytoplasm alone.

The phosphoglyceryl kinase which catalyses the accompanying reaction

84) $COO\textcircled{P}CHOH\ CH_2O\textcircled{P} + ADP \rightleftharpoons COOH\ CHOH\ CH_2O\textcircled{P} + ATP$

appears to be the same enzyme in both carbon assimilation and glycolysis and is found in both chloroplasts and cytoplasm. The enzymes triosephosphate isomerase and aldolase, which are also involved in the carbon pathways of both photosynthetic carbon fixation and glycolysis, have also been shown to occur in chloroplasts and cytoplasm of leaves, algae and *Euglena*. Non-aqueous methods of extraction have been mainly depended on, since the proteins of the chloroplast stroma are readily lost to aqueous solvents.

The CO_2-acceptor in chloroplasts is ribulose-1,5-diphosphate. The reactions leading to 3-phosphoglyceric acid have much in common with the pentose phosphate pathway (p. 68) and might be regarded as its reductive equivalent. Many of the enzymes involved in the two processes are apparently identical; but the one process is linked to the chloroplast stroma, the other to soluble cytoplasm.

While the membrane-bound electron transport pathways of photosynthesis and respiration seem to be firmly separated in different organelles, the same cannot be said so confidently of the other stages of the processes. Since the enzymes common to carbon dioxide fixation and to glycolysis or the pentose phosphate pathway are so easily removed from chloroplasts by aqueous media, it does not even seem certain that some transfers may not occur between stroma and soluble cytoplasm in the living cell. What is more certain is that transfers of intermediate products can occur. There is evidence that 3-phosphoglyceric acid produced in algal glycolysis is used directly in photosynthesis. Conversely, isolated chloroplasts yield glyceraldehyde-3-phosphate and dihydroxyacetone phosphate to the medium when illuminated. It is improbable that exchanges of NAD or NADP occur between chloroplasts, mitochondria and the soluble cytoplasm; but AMP, ADP and ATP often appear to be more mobile in the cell. Photosynthesis and respiration both require ADP; and, as photosynthetic rates may be an order of magnitude faster than respiratory ones in the same cell, it might be expected that they would suppress respiration entirely. Attempts to measure respiration rates of green cells during illumination by means of labelled oxygen ($^{18}O_2$) do not usually suggest that this occurs. It may therefore be that chloroplast membranes, at least, do not readily allow the passage of the adenosine phosphates, and keep separated the respiratory and photosynthetic supplies. The only recorded cases of complete suppression of respiration by photosynthesis are for *Anabaena*, a blue-green alga where the photosynthetic and respiratory sites are both located on the lamellae of the outer cytoplasm (Plate II), and even here total suppression is not the rule. When ^{14}C-labelled substrates are fed to the green alga *Scenedesmus*, labelling still appears on citric acid cycle intermediates without much difference between light and dark.

The photorespiration that leads to an enhanced CO_2-output by some green cells in light has already been described on p. 99.

Nuclei: glycolysis

Nuclei are large consumers of energy provided by ATP, and in a young cell may be a large part of the cell mass. They are the site of much protein as well as DNA and RNA formation at all stages. The source of the ATP utilised is at least to some extent within the nucleus itself. Glycolysis has been shown to occur in nuclei extracted from a wide variety of cells from both animals and plants. Nuclei extracted in aqueous media usually glycolyse fructose-1,6-diphosphate, but not glucose; whereas those extracted in non-aqueous solvents may be able to perform a complete conversion from glucose to lactic acid. Individual glycolytic reactions and enzymes and NAD have also been identified showing that the stages are the same as those occurring in the soluble cytoplasm. The responsible enzymes appear to be situated mainly in the nuclear sap.

The occurrence of the citric acid cycle of reactions also seems possible in nuclei. Pyruvate-3 ^{14}C can be oxidised to $^{14}CO_2$, and some of the enzymes and intermediate products of the cycle are found in nuclei extracted in non-aqueous media.

The occurrence of oxidative phosphorylation in nuclei is more dubious. It has been reported that the nuclei of calf thymus require oxygen to phosphorylate AMP and ADP to ATP. The system involved differs from the mitochondrial system and its components have not been identified. It has proved difficult to obtain corresponding results with nuclei from liver, kidney and pancreas. The oxygen consumption recorded for thymus nuclei was very small, $Q_{O_2} = 0.5$ with much of it due to cytoplasmic contamination (cf., Table 3, p. 17).

Transport of ATP

There is general agreement that nuclei do not contain the constituents of the electron transport system found in mitochondria. Apparent traces occasionally reported have been attributed to mitochondrial contamination. Sonicated wheat nuclei revealed no cytochrome $a + a_3$ spectroscopically at a total N_2 concentration 20-fold that required with mitochondria.

It therefore seems doubtful whether the heavy ATP demands of developing nuclei can be met by their own phosphorylations. Alternatively they might receive additional supplies from the mitochondria. Nuclei extracted into aqueous media from wheat embryos have been shown to synthesise protein in significant amounts when supplied externally with 0·01 M ATP. This needed to be present only at low concentration (0·001 M), or could be substituted with a similar concentration of ADP, if mitochondria were added. Protein synthesis was stopped in the nuclei when the mitochondrial respiration was inhibited with CO or uncoupled from ATP formation with DNP. It continued when the nuclei and mitochondria were separated by a membrane permeable to ATP, enforcing a diffusion path very long in terms of cell dimensions. The membranes of these nuclei were ruptured during extraction (Plate VIIIa). In the living cell, nucleoplasm and cytoplasm are in contact at the numerous 'pores' which bridge the perinuclear space. Up to the present no method other than diffusion has been demonstrated for the movement of ATP in the cytoplasm. The mitochondria are themselves mobile in the cytoplasm and are often said to cluster round points of ATP consumption, such as the bases of flagella; there has even been described a shuttling of mitochondria between cell surface and nuclei. It is probably wise not to place too much significance on such observations until an actual transfer of ATP *in vivo* has been demonstrated.

Conclusion

Summarising present knowledge of the relation between the metabolic functions and substructure of cells it is evident that the connection is far from being a simple one. Principal functions of a cell are not allotted to specific parts or organelles; but may overlap and repeat with varying degrees of efficiency in different situations. It may not be too much to say that four principal phases; i.e. the mitochondria; the nucleus; chloroplasts where present; and the cytoplasm consisting of the endoplasmic reticulum bathed in a soluble matrix are each in themselves capable of all the essential functions of energy capture, storage and utilisation. Each of them, however, shows special efficiency in one or other branch; such as anaerobic oxidation in the soluble phase, reductive energy fixation in chloroplasts and so on. The entire cell as a

consortium of these parts is able to employ all their special skills to give a high
overall efficiency and great versatility of operation. The origin and evolution
of such a system is obviously a subject of the greatest interest; but too complex
and speculative to be considered here.

Further Reading

De Duve, C., Wattiaux, R. and Baudhuin, P. (1962) Distribution of enzymes
between subcellular fractions in animal tissues. *Adv. in Enzymol.*, **24**, 291–358.
Green, D. E. (1959) Mitochondrial structure and function. In *Subcellular Particles.*
Ronald Press, New York.
James, W. O. and Richens, A. M. (1962) Energy transport from mitochondria to
nuclei. *Proc. Roy. Soc.*, **B 157**, 149–59.
Lehninger, A. L. (1959) Metabolic interactions in cell structures. In *Developmental
Cytology.* Ronald Press, New York.
Marchant, R. and Smith, D. G. (1968) Membranous structures in yeast. *Biol. Rev.*,
43, 459–80.
Roodyn, D. B. (editor) (1967) *Enzyme Cytology.* Academic Press, New York.

Recent detailed work on the connexions between the electron transport particle,
phosphorylation and mitochondrial structure is not described in the text. It is
surveyed in such publications as:

Hall, D. O. and Palmer, J. M. (1969) Mitochondrial research today. *Nature*, **221**,
717–23.
Pullman, M. E. and Schatz, G. (1967) Mitochondrial oxidations and energy coupling.
Ann. Rev. Biochem. **36**, 539–610.
Racker, E. (1970) *Membranes of the mitochondria and chloroplast.* Van Nostrand,
New York.
Racker, E. (1970) The two faces of the inner mitochondrial membrane. *Essays
in Biochemistry*, **6**, 1–22, Academic Press, London.
Wrigglesworth, J. M., Packer, L. and Branton, D. (1970). Organisation of mito-
chondrial structure as revealed by freeze-etching. *Biochem. Biophys. Acta.*,
205, 125–35.

Index

Italic entries include an illustration or a formula.